Engineering Mechanics for Structures

Engineering Mechanics for Structures

Finn Gibson

STATES
ACADEMIC PRESS
www.statesacademicpress.com

States Academic Press,
109 South 5th Street,
Brooklyn, NY 11249, USA

Visit us on the World Wide Web at:
www.statesacademicpress.com

ISBN: 978-1-63989-713-1

Cataloging-in-publication Data

Engineering mechanics for structures / Finn Gibson.
p. cm.
Includes bibliographical references and index.
ISBN 978-1-63989-713-1
1. Structural analysis (Engineering). 2. Mechanics, Applied. 3. Structural engineering.
4. Engineering. I. Gibson, Finn.

TA645 .E54 2023
624.171--dc23

Table of Contents

Preface

The main aim of this book is to educate learners and enhance their research focus by presenting diverse topics covering this vast field. This is an advanced book which compiles significant studies by distinguished experts in the area of analysis. This book addresses successive solutions to the challenges arising in the area of application, along with it; the book provides scope for future developments.

Structural mechanics is usually studied under the field of applied mechanics. It is the methodological investigation of the deformations, deflections, and internal forces or stresses (stress equivalents) within structures. Structural analysis and design plays an instrumental role in generating a structure that is capable of resisting all applied loads without failure during its intended life. Mechanics for structures is a field of study that examines the behavior of structures under mechanical loads, such as bending of a beam, buckling of a column, torsion of a shaft, deflection of a thin shell, and vibration of a bridge. There are three basic approaches to the mechanical structures analysis, namely, the energy methods, flexibility method, and the direct stiffness method. These methods later developed into finite element method and the plastic analysis approach. The book studies and analyses the most significant concepts and aspects of engineering mechanics for structures. It will serve as an essential guide for both undergraduate and graduate students of civil engineering and engineering mechanics.

It was a great honour to edit this book, though there were challenges, as it involved a lot of communication and networking between me and the editorial team. However, the end result was this all-inclusive book covering diverse themes in the field.

Finally, it is important to acknowledge the efforts of the contributors for their excellent chapters, through which a wide variety of issues have been addressed. I would also like to thank my colleagues for their valuable feedback during the making of this book.

Finn Gibson

1

Introduction to Civil Engineering and Engineering Mechanics

1.1 Scope of Different Fields of Civil Engineering

Introduction to Civil Engineering

Civil engineering is the technology that includes numerous other disciplines which produce useful facilities for the human beings including water disposal, dams, roads and other facilities that are used in our daily life. Civil engineering is progressing at a fast pace as other technologies.

Works by Civil Engineering

Civil engineering is considered as the first discipline of the various branches of engineering after the military engineering and it includes the planning, designing, construction and maintenance of the infrastructure.

These works includes the buildings, dams, roads, bridges, canals, water supply and many other facilities that affect the life of human beings.

Civil engineering is intimately associated with the public and private sectors including the individual homeowners and the international enterprises. It is one of the ancient engineering achievements and oldest engineering professions.

Work By Civil Engineering.

Civil Engineering in Daily Life

Civil engineering has a significant role in the life of every human being, though we may not sense its importance in our day to day activities.

The function of civil engineering commences with the start of the day when we take a shower, since the water is delivered with the help of the water supply system including a well-designed network of the pipes, water treatment plant and other numerous associated services.

The network of roads on which we drive while proceeding to work or school, the huge structural bridges we come across and the tall buildings where we work, all have been designed and constructed by the civil engineers.

Even the benefits of electricity we use are available through the contribution of the civil engineers who constructed the towers for the transmission lines.

In fact, no sphere of life in this earth may be identified, that does not include the contributions of civil engineering.

Thus, the importance of civil engineering may be determined according to its usage in our daily life.

Falkirk - Wheel large.

Future of Civil Engineering

Civil engineering use the technical information obtained from numerous other sciences and with the advancement in all the types of technologies, the civil engineering has also benefited tremendously.

The future of civil engineering is expected to be revolutionized by the modern technologies including the design software, GPS, GIS systems and other latest technical expertise in various fields.

Technology will continue to make the important changes in the application of civil engineering including the rapid progress in the use of 3-D and 4-D design tools.

Scope of Different Fields of Civil Engineering

Due to the increase in the scope of civil engineering with the passage of time, it has now got divided into many branches of study.

Some of the significant ones include geotechnical engineering, environmental engineering, and structural engineering, and hydraulic engineering, transportation engineering and much more important areas of study.

Engineers are being employed by a wide range of companies in the US, from small startup businesses focused on a new invention idea to large-scale companies that work on huge contracts.

The engineers from different fields constantly work together to create successful products.

While considering the design and manufacture of an aircraft, for example, the workforce behind the development will include analysis engineers evaluating the strength of landing gear developed by design engineers, aeronautical engineers optimizing airflow paths, electronics engineers developing the pilot controls and wiring methods, computer engineers programming the aircraft operation systems, including everything from the autopilot system to the cabin crew call system and ergonomic engineers designing comfortable seating.

Apart from structures on land and general transportation systems, civil engineers are also responsible for building good transportation systems for flow of water, i.e. the water distribution systems.

The main activities in this undertaking are designing the pipelines for flow of water, dams, canals, drainage facilities etc. Dams are a major source for non-conventional energy and are hence in high demand today.

While designing these structures, the civil engineers must take into account the various properties of fluids to calculate the forces acting at different points.

1.1.1 Surveying

Surveying is defined as the process of measuring horizontal distances, vertical distances and included angles to determine the location of points on, above or below the earth surfaces.

The term surveying is the representation of surface features in a horizontal plane. The process of determining the relative heights in the vertical plane is referred as levelling.

Objectives of Surveying

The data obtained by surveying are used to prepare the plan or map showing the ground features. When the area surveyed is small and the scale to which its result plotted is large, then it is known as Plan.

When the area surveyed is large and the scale to which its result plotted is small, then it is known as a Map. Setting out of any engineering work like buildings, roads, railway tracks, bridges and dams involves surveying.

Example: A plan of a building, a map of India.

1.1.2 Building Materials and Construction Technology

Stones, bricks, cement, lime and timber are the traditional materials used for civil engineering constructions for several centuries.

Stones

Stone is a 'naturally available building material' which has been used from the early age of civilization. It is available in the form of rocks, which is cut to required size and shape and used as building block. It has been used to construct small residential buildings to large palaces and temples all over the world. Red Fort, Taj Mahal, Vidhan Sabha at Bangalore and several palaces of medieval age all over India are the famous stone buildings.

Brick

Brick is obtained by moulding good clay into a block, which is dried and then burnt. This is the oldest building block to replace stone. Manufacture of brick started with hand moulding, sun drying and burning in clamps. A considerable amount of technological development has taken place with better knowledge about properties of raw materials, better machineries and improved techniques of moulding ,drying and burning.

The size of the bricks are of 90 mm × 90 mm × 90 mm and 190 mm × 90 mm × 40 mm. With mortar joints, the size of these bricks are taken as 200 mm × 100 mm × 100 mm and 200 mm × 100 mm:

- Building Bricks: These bricks are used for the construction of walls.
- Paving Bricks: These are vitrified bricks and are used as pavers.
- Fire Bricks: These bricks are particularly made to withstand furnace temperature. Silica bricks belong to this category.
- Special Bricks: These bricks are different from the commonly used building bricks with respect to their shape and the purpose for which they are made.

Tests on Bricks

The following laboratory tests may be conducted on the bricks to find their suitability:

- Crushing strength.
- Absorption.
- Shape and size.
- Efflorescence.

Uses of Bricks

Bricks are used in the following civil works:

- As building blocks.
- For lining of ovens, furnaces and chimneys.
- For protecting steel columns from fire.
- As aggregates in providing water proofing to R.C.C. roofs.
- For pavers for footpaths and cycle tracks.
- For lining sewer lines.

Sand

Sand is a natural product which is obtained as river sand, nalla sand and pit sand. However sea sand should not be used for the following reasons:

- It contains salt and hence, structure will remain damp. The mortar is affected by efflorescence and blisters appear.
- It contains shells and other organic matter, which decompose after some time, reducing the life of the mortar.

Sand may be obtained artificially by crushing hard stones. Usually artificial sand is obtained as a by-product while crushing stones to get jelly (coarse aggregate).

Cement

Cement is a commonly used binding material in the construction. It is obtained by burning a mixture of calcarious (calcium) and argillaceous (clay) material at a very high temperature and then grinding the clinker so produced to a fine powder. It was first produced by a mason Joseph Aspdin in England in 1924. He patented it as portland cement.

Uses of Cement

Cement is used widely for the construction of various structures. Some of them are listed below:

- Cement slurry is used for filling cracks in concrete structures.
- Cement mortar is used for masonry work, plastering and pointing.
- Cement concrete is used for the construction of various structures like buildings, bridges. water tanks, tunnels, docks, harbours, etc.
- Cement is used to manufacture lamp posts, telephone posts, railway sleepers, piles, etc.
- For manufacturing cement pipes, garden seats, dust bins, flower pots etc. cement is commonly used.
- It is useful for the construction of roads, footpaths, courts for various sports, etc.

Plain Concrete

Plain concrete, commonly known as concrete, is an intimate mixture of binding material, fine aggregate, coarse aggregate and water. This can be easily moulded to desired shape and size before it looses plasticity and hardens.

Plain concrete is strong in compression but very weak in tension. The tensile property is introduced in concrete by inducting different materials and this attempt has given rise to RCC, RBC, PSC, FRC, cellular concrete and Ferro cement.

Major ingredients of concrete are:

- Binding material (like cement, lime, polymer).
- Fine aggregate (sand).
- Coarse aggregates (crushed stone, jelly).
- Water.

Reinforced Cement Concrete

RCC (Reinforced Cement Concrete) is the combination of using steel and concrete instead of using only concrete to offset some limitations. Concrete is weak in tensile stress with compared to its compressive stress.

To offset this limitation, steel reinforcement is used in the concrete at the place where the section is subjected to the tensile stress. Steel is very strong in its tensile stress.

The reinforcement is usually round in shape with approximate surface deformation is placed in the form in advance of the concrete. When the reinforcement is surrounded by the hardened concrete mass, this forms an integral part of the member.

The resultant combination of two materials are known as reinforced concrete. In this case the cross-sectional area of the beam or other flexural member is greatly reduced.

Reinforced concrete (RCC).

Steel Sections

Steel is the extensively used building material. The following three varieties of steel are extensively used:

- Mild steel.
- High carbon steel.
- High tensile steel.

1. Mild Steel

It contains a maximum of 0.25% carbon, 0.055% of sulphur and 0.55% of phosphorus.

Uses of Mild Steel:

- Round bars are extensively used as reinforcement in R.C.C. works.
- Rolled sections like I, T, L, C, plates, etc. are used to build steel columns, beams, trusses, etc.
- Tubular sections are used as poles and members of trusses.
- Plain and corrugated mild steel are used as roofing materials.
- Mild steel sections are used in making parts of many types of machinery.

2. High Carbon Steel

The carbon contents in this steel is 0.7% to 1.5%.

Uses of High Carbon Steel:

- It is used for making tools such as drills, files, chisels.
- Many machine parts are made with high carbon steel since it is capable of withstanding shocks and vibrations.

3. High Tensile Steel

It contains 0.8% carbon and 0.6% manganese. The strength of this steel is quite high. High tensile steel wires are used in prestressed concrete works.

Timber

Timber is the wood prepared for use in building and carpentry.

Plywood

Plywood is a type of sheet material which is manufactured from the thin layers or "plies" of wood veneer that are glued together with the adjacent layers having wood grain rotated at about 90 degrees to one another.

Plywood is an engineered wood from the family of manufactured boards which includes medium-density fibre board and particle board.

Paints

Paint is any liquid, liquefiable or mastic composition that after application to a substrate in a thin layer converts to a solid film. It is most commonly used to protect, color or provide texture to objects.

Varnish

Varnish is the transparent, hard, protective finish or film primarily used in wood finishing but also for other materials. Varnish is traditionally a combination of a drying oil, a resin and a thinner or solvent.

Varnish finishes are usually glossy but may be designed to produce satin or semi-gloss sheens by the addition of "flatting" agents. It has little or no color is transparent and has no added pigment as opposed to paints or wood stain, which contain pigment and generally range from opaque to translucent.

Varnishes are also applied over wood stains as a final step to achieve a film for gloss and protection. Some products are marketed as a combined stain and varnish.

Construction Technology

A suitable environment is created by constructing a building. The building technology covers the planning of different units of a building to provide a suitable environment for the activities designed for a building.

The building technology also covers the maintenance and repairs of the buildings and their safe demolition when they become too old to be used further.

The buildings are classified according to functions such as:

- Residential buildings.
- Public buildings.
- Industrial buildings.
- Commercial buildings.
- Recreation buildings.
- Hospital buildings.
- Educational buildings.
- Storage. i.e. warehouses, etc.
- Special purpose buildings, non-conventional buildings.

Building technology deals with the analysis and design of substructures as well as superstructures of the buildings. It includes the study of different construction materials in respect of their properties and construction techniques. Some of the building materials are metals, timber, concrete, bituminous materials, soil, bricks, polymers and plastics, etc.

Construction Technology and its Management

The scope of 'construction' is more comprehensive here than that is in building technology. Based on the sound principles of soil mechanics, foundations to non-conventional structures are covered under construction technology.

It is comprised of different techniques of construction for different materials under different site conditions. The management or organization of men (labour), materials, methods in relation to site, money and time is the backbone of construction management.

It involves almost every branch of engineering, commerce and economics, and its ultimate aim is to 'achieve the desired construction in the most economic way'. A clear knowledge of the following points is necessary for reliable construction and its management.

1.1.3 Geotechnical Engineering

Geotechnical engineering is the civil engineering discipline that deals with study of soil properties and engineering behaviour of soil under the action of particular loads and moisture content. It includes soil mechanics, some aspects of geology and foundation engineering.

Physical properties of soil has great impact on stability and safety of structures. Particle size, chemical composition, moisture content and swelling-shrinking of soil are the factors that control bearing capacity of soil which must be considered for design of foundations, roads, buildings, etc., In general, this discipline is very close to structural engineering.

Geotechnical engineering.

Applications

- Sub-soil exploration i.e. determination of important physical and engineering properties of soil lying underneath.
- Design and construction of foundations for water structures, buildings and machines.
- Effective, efficient and economical type of foundation for load transfer on wider area.
- Estimation of bearing opacity (B.C.) of soil and improving the B.C.

- Repair/maintenance of foundations (if possible) to avoid failure and/or settlement.
- Design of retaining walls, earthen dams/embankment.
- Ascertaining stability of ground slopes and/or improving the stability to avoid/control landslides.
- Improvement in foundation design, type and construction for specific requirements like atomic power plant, airports, docks/harbours/ports.
- Study of sub-soil layers and soil profile for preparing soil maps.
- Advanced studies and applications such as estimation of depth of water table, ground water, seismic activities, fault/plate movements at different depths below ground level.
- Design and construction of transportation routes and allied structures such as bridge, shafts, tunnel, etc.
- Water seepage analysis for dams.

1.1.4 Structural Engineering

Structural engineering is a branch of civil engineering that includes safe and economical design of structures and structural members as well as connections such as rivets, bolts, welds, keys, etc.

For a given loading, suitable cross section of members can be determined.

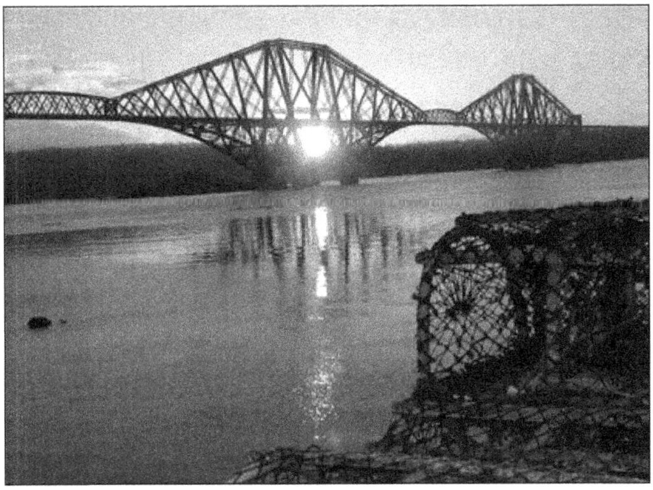

Forth bridge.

Structural engineering further includes engineering mechanics, strength of materials, theory of structures, design of steel and Reinforced Cement Concrete (R.C.C) structures, etc. For the given loading, the conventional materials as well as alternate/modern materials are used to get minimum possible size of structural members with adequate factor of safety viz. prestressed concrete.

Applications

- Design and erection/construction of structural members as well as structures, connectors/fixtures with adequate factory of safety and economy.
- Design of superstructure and substructure of a building or factory shed, etc.
- Investigation of failure of a member or structure for assigning responsibility and for avoiding repetition of the mistakes/causes of failure.
- Design of steel components and structures including water tanks.

1.1.5 Hydraulics, Water Resources and Irrigation Engineering

Hydraulics

Hydraulics is the science of transmitting force and/or motion through the medium of a confined liquid. In a hydraulic device, power is transmitted by pushing a confined liquid.

The transfer of energy takes place, because a quantity of liquid is subjected to pressure. To operate liquid-powered systems, the operator should have a knowledge of the basic nature of liquids.

Water Resources Engineering

Water resources engineering concerns with the management of quantity and quality of the water in the underground and above ground water resources such as rivers, lakes and streams.

Water resource engineering.

Geographical areas are analyzed to forecast the amount of water that will flow into and out of the water source. Fields of hydrology, geology and environmental science are included in this discipline of civil engineering.

Irrigation Engineering

It is another very important discipline of civil engineering that deals with tapping or storage of water and supplying water either for crop cultivation or for drinking and other domestic/industrial uses.

Two important basic human needs viz. water and food are taken care of by irrigation engineering directly and indirectly. Water as an important and scarce natural resource is in the form of surface water (run off after rainfall) and sub-surface water (ground water) which is stored in dams and reservoirs or tapped from wells (open wells, tube wells, etc.)

An irrigation project may be multipurpose. i.e. for water supply as well as for hydropower generation. Irrigation engineering thus includes estimation of quality and quantity of water available, its storage and distribution through open canals or pipe-networks either under gravity or by pumping.

Agriculture mainly depends on irrigation since all areas do not have sufficient rainfall and water is made available in non-rainy seasons also for cropping. Irrigation engineering under takes flood control also.

Application

- Estimation of quality and quantity of surface and subsurface water available at given places.
- Water supply schemes for cropping, drinking and other purposes.
- Calculations of reservoir capacity, catchment area, command area, etc.
- Design and construction of reservoir and dams.
- Design and construction of laying canals, pipes, pumps and allied structures such as spillways, gates, weirs-notches, etc.
- Flood control arrangements, diversion works, etc.
- Estimation of water requirement for cropping, so that most efficient and economical use of water is made.
- Recommending or providing non-conventional effective techniques such as drip irrigation, sprinkler irrigation, etc.

- Study and recommendation of cropping pattern or crop rotation for efficient use of water and higher yield such as inter-cropping, alternate cropping, etc.
- Avoiding water logged areas and salination of soil and also control soil erosion.

1.1.6 Transportation Engineering

Transportation engineering is the discipline that deals with study of present transportation systems and their improvement for safe, economical and efficient (in less time) conveyance of materials/goods/finished products as well as human beings and animals.

It Includes design, construction and management of roads, railways, navigation and air-routes. Allied constructions such as tunnels, culverts, bridges, aqueducts are also covered in the sub-disciplines such as bridge engineering, highway engineering, tunnelling, etc.

Traffic management including traffic signals, number of lanes, parking facilities and curves are also a part of transportation engineering.

Applications

- Design and construction of different types of roads.
- Traffic management and parking facilities.
- Design and provision of curves and allied structures such as bridge, tunnel, culvert and ghat-roads.
- Survey, design and provision of different modes of transportation. viz. airports, roads, railways, ports and harbours, etc.
- Use modem techniques of management to ensure rapid transportation of people and goods/raw materials/agricultural produce with sufficient convenience, comfort, economy and safety.
- Avoid heavy traffic through cities or villages by providing bye-pass/diversion roads and expressway.
- Help the economic growth of regions and country through fast transportation system.
- Provide durable and strong as well as safe modes of transport and repair/maintenance with least possible delays and inconvenience.

1.1.7 Environmental Engineering

Environmental engineering discipline deals with study of the natural environment/ecosystems, inter-relation between biotic and abiotic factors, safety of people against different types of pollution and treatment-disposal of wastes.

It includes water supply engineering sanitary engineering and environmental studies. Considering the increase in population and rapid rate of urbanization, all types of pollution have increased and health of people is at stake by and large.

Hence environmental impact assessment for industries, factories and control/eradication/prevention of pollution have become very essential areas for study and research/development.

Environmental engineering.

Applications

- Measurement of pollutants as regards air, water, radioactive and noise pollution.
- Water treatment for supplying potable (drinking) water.
- Waste water treatment- design, construction and repair/maintenance of treatment plant.
- Research and development for recycling or reusing the mass energy from wastes.
- Determining or fixing standards for effluents.
- Monitoring, control and/or prevention of different types of pollution.
- Environmental Impact Assessment/analysis for factories and industries.

1.2 Infrastructure

Civil infrastructure systems involves the analysis, design and management of infrastructure, supporting human activities, including water and wastewater, oil and gas, electric power, communications, transportation and the collections of buildings that make up rural and urban communities.

These networks deliver essential services, support social interactions and economic development and provides shelter. They are society's lifelines.

The field of civil infrastructure systems builds on and extends traditional civil engineering areas. Rather than focus on individual structural components or structures, civil infrastructure systems emphasizes on how different structures behave together as a system that serves the community's needs.

Problems in this field typically involves a great deal of uncertainty, multiple and competing objectives and sometimes conflicting and numerous constituencies. They are often spatial and dynamic.

The technical aspects of infrastructure engineering must be understood in the economic, social, political and cultural context in which they exist and must be considered over a long-time horizon that includes not just design and construction, but also performance, maintenance and operations in natural disasters and other extreme events and destruction as well.

1.2.1 Types of Infrastructure

Infrastructure systems include both fixed assets and control systems and software required to operate, manage and monitor the systems, as well as any accessory buildings, plants or vehicles that are an essential part of the system. The various types of infrastructures are listed below,

Transportation Infrastructure

- Road and highway networks, including structures (bridges, tunnels, culverts, retaining walls), signals and markings, electrical systems and edge treatments (curbs, landscaping, sidewalks).
- Railways, including structures, terminal facilities (train stations, rail yards), level crossings, signalling and communications systems.
- Canals and navigable waterways requiring continuous maintenance (dredging, etc.).
- Airports, including air navigational systems.
- Seaports and lighthouses.
- Mass transit systems (commuter rail systems, tramways, subways, trolleys and bus terminals).

- Bicycle paths and pedestrian walkways.

Financial Infrastructure

- Banking system.
- Exchanges.
- Money supply.
- Financial regulations.

Energy Infrastructure

- Electrical power network including electric grid, generation plants, substations and local distribution.
- Natural gas pipelines, storage and distribution terminals as well as the local distribution network.
- Petroleum pipelines, including the associated storage and distribution terminals.
- Steam or hot water production and distribution networks for direct heating systems.

Water Management Infrastructure

- Drinking water supply, including the system of pipes, reservoirs, valves, pumps, filtration and treatment equipment and meters. It also includes buildings and structures to house the equipment, used for collection, treatment and distribution of drinking water.
- Sewage collection and disposal.
- Drainage systems (storm sewers, ditches, etc.)
- Major irrigation systems.
- Major flood control systems (dikes, levees, major pumping stations and floodgates).

Communication Infrastructure

- Communication software.
- Social network services.

- Postal services.
- Telephone networks, including switching systems.
- Mobile phone networks.
- Cable television networks including receiving stations and cable distribution networks.
- Internet backbone, including high speed data cables, routers and servers as well as the protocols and other basic software required for the system to function.
- Communication satellites.
- Undersea cables.
- Major Private, government or dedicated telecommunication networks, such as those used for internal communication and monitoring by major infrastructure companies, by governments, by the military or by emergency services.
- Pneumatic tube mail distribution networks.

1.2.2 Role of Civil Engineer in the Infrastructural Development

A civil engineer has to conceive, plan, estimate, get the approval, create and maintain all civil engineering activities.

A civil engineer has a very important role in development of the following infrastructure:

- Measure and map the earth's surface.
- Plan for the new townships and extension of existing towns.
- Construct suitable structures for rural and urban areas.
- Build tanks and dams to exploit the water resources.
- Build river navigation and flood control projects.
- Construct canals and distributaries to take water to agricultural fields.
- Purify and supply water to the required areas such as houses, schools and offices.
- Provide and maintain the communication systems such as roads, railways, harbours and airports.
- Devise systems for control and efficient flow of traffic.

- Provide and maintain solid and waste water disposal system.
- Monitor land, water and air pollution and take measures to control them. Fast growing industrialization has put heavy responsibilities on the civil engineers to preserve and protect environment.

1.2.3 Effect of the Infrastructural Facilities on Socio-economic Development of a Country

Civil engineering activities in the infrastructural development include:

- Good planning of towns and extension areas in cities. Each extension area should be self-sufficient in accommodating educational institutions, offices, hospitals, markets, recreational facilities and residential accommodation.
- Assured water supply.
- A good drainage system.
- Pollution free environmental conditions.
- A well planned and built network of roads and road crossings.
- Railways connecting all the important cities and towns.
- Airports and harbours of national and international standards.

Infrastructure also involves electricity supply, without assured electric supply no city or town can develop. Educational facility also forms pan of infrastructure. Good primary and secondary schools nearby residential areas are desirable. Collegiate and professional educations also form part of infrastructure of a city.

Further, good health care facility is very essential. Primary health centres, specialized hospitals and doctors add to the desirable infrastructure facility.

Effects of infrastructure facilities are:

- Connecting production centres to marketing places minimizes exploitation of producers by middlemen. Imports and exports become easy.
- Improved irrigation facility enhances the agricultural products and hence, producers as well as consumers are benefited.
- Infrastructural facility develops scope for a number of industries and it creates job opportunities.
- Improved education and healthcare adds to skilled and healthy workforce. Quality of life of the people is improved.

- Utilization of manpower for the benefit of mankind brings down anti-social activities.
- In case of natural calamities, assistance can be easily extended to the affected areas and misery of affected people gets minimized.
- Infrastructural facility improves defence system and peace exists in the country.
- Improved economical power of the country brings a respectable status in the world. It is realized that a government should not involve itself in production and distribution, but should develop infrastructure to create an atmosphere for economical development.

1.3 Roads: Classification and their Functions

The roads are named according to the type of jurisdiction, constructions and important function etc.:

- Names like metalled roads, earth, concrete roads and asphalt roads, indicates the type of constructions.
- Names like local roads, state highways, district roads, national highways indicates their jurisdiction.
- Names like rectangular roads, ring roads, radial and circular roads and diagonal roads indicate their geometric shape.
- Names like Avenue, promenade, boulevards and parkways indicate their dominant function.

Types of Roads

Various criteria may be used for classifying roads.

Based upon the usage of roads during rainy season they may be classified as:

- All weather roads.
- Fair weather roads.

All weather roads are the roads that are not flooded during rainy seasons except to a small extent at river crossing for a small length. In fair weather roads overflow of streams across the road is permitted during monsoon season.

Depending on the type of pavement surfaces provided, the roads may be classified as:

- Surfaced roads.

- Unsurfaced roads.

Surfaced roads are provided with a concrete or bituminous surface while unsurfaced roads may be mud roads or water bound macadam layer roads.

The Nagpur road plan categorized the roads in India into following five categories:

- National Highways (NH).
- State Highways (SH).
- Major District Roads (MDR).
- Minor or Other District Roads (ODR).
- Village Roads (VR).

National Highways (NH)

These are the roads that connects important cities, towns, ports, etc., of different states. They may connect even the neighbouring countries also. The National Highways have two-lane traffic at least 8 m wide with at least 2 m wide shoulders on each side.

The construction and maintenance of these roads is taken care by the Central Government agencies like Military Engineering Service (MES) or Central PWD. The National Highways are assigned the respective numbers.

The highway connecting Delhi-Ambala-Amritsar is denoted as NH-1. The National Highway connecting Pune-Bangalore-Chennai is known as as NH-4. The west-coast highway that connects Bombay to Kanyakumari is known as NH-17.

State Highways (SH)

These are important roads of a particular state connecting the important cities and district headquarters. They connect important cities to national highways. They are maintained by the State Public Works Departments and Central Government gives grants for the construction and development of these roads.

These highways also have 8 m carriage way and 2 m wide shoulders on each side. The design speed and the design specifications of State Highways are same as those for National Highways.

Major District Roads (MDR)

These are the roads within a district connecting market and production areas to National or State Highways or railway stations. The MDR has lower speed and geometric design specifications than for NH or SH.

Minor or Other District Roads (ODR)

These roads connect rural areas of production to market centres, taluk centres or other main roads. These roads have lower design specifications than MDR. These roads are looked after by district authorities with the help of State Government Departments.

Village Roads (VR)

The roads connecting villages or group of villages with each other or the roads of higher category. The local district boards are responsible for the construction and maintenance of these roads. These roads are generally unmetalled.

After the third road development plan (Lucknow road plan) the roads in the country are categorized into three classes, viz:

- Primary system.
- Secondary system.
- Tertiary system or rural roads.

Primary system consists of two classes:

- Expressways.
- National Highways (NH).

Expressways are superior to National Highways and are provided wherever traffic volume is very high. They have superior facilities and design speed. No cross-traffic is allowed on expressways.

They are provided with central separator for the traffic in opposite direction. They are fenced so that animals do not enter. Controlled access is provided to other roads, towns and cities.

Only fast moving vehicles are permitted. Expressways may be owned by State or Central Government. Golden Quadrilateral connecting Delhi, Mumbai, Chennai and Kolkata is owned by the Central Government.

This line passes through Belgaum, Chitradurga, Dharwad-Hubli, Davangere, Tunkur and Bangalore in Karnataka, Bangalore–Mysore Infrastructure Corridor expressway, which is under construction, is owned by the State Government.

The construction works have been undertaken on the basis of Build–own-Operate–Transfer (BOOT) by private parties on contract assigned by the respective governments.

The secondary system consists of two types of roads, namely, State Highways and Major District Roads.

The third category of roads consist of Village Roads and Other District Roads.

Urban roads form a separate class of roads, which are taken care by municipal corporations or municipalities. These roads may consist of the following:

- Arterial roads.
- Collector streets.
- Expressways.
- Sub-arterial roads.
- Local streets.

Local streets are about to private properties like houses and shops. They are connected to collector streets.

Components of Roads

All roads consist of the following components:

- Pavement or carriageways.
- Shoulder.

Pavement or Carriageways

This is the width of the road which is designed to handle volume of expected traffic. As per Indian Road Congress specification, the maximum width of vehicle is 2.44m.

A side margin of 0.68 m is needed for safe driving of the vehicle. Hence for a single lane road carriageway width works out to be 3.8m. For road pavements having two or more lanes, the width of 3.5m per lane is considered sufficient.

Number of lanes needed for a road is decided by the volume of traffic to be handled and also financial considerations.

The cross-section of carriageway comprises of the following components:

- Subsoil.
- Subgrade.
- Base.
- Surfacing.

Subsoil is prepared which will take the load of the road. It is prepared by properly compacting the natural soil.

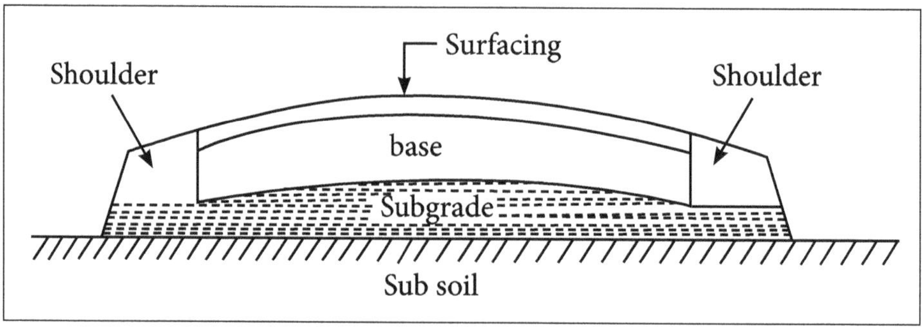

Cross-section of carriageway.

Subgrade gives support to the road structure. It should remain dry and stable throughout. Considerable attention is to be given for laying proper subgrade to get stable road surface. The subgrade soil mainly consists of disintegrated rocks like clay, silt, gravel and sand.

The desirable properties of subgrade soil are:

- Incompressibility.
- Stability.
- Permanency of strength.
- Minimum change in volume.
- Ease of compaction.
- Good drainage.

The base might consist of two layers, top layer being called as base and bottom layer as sub-base.

Base course and sub-base course distribute the load through a finite thickness. The sub-base layer is made with stabilized soil or selected granular soil, boulders or bricks.

However, it would be better if graded aggregates with soil are used instead of boulders. Base course is provided with broken stone aggregates.

Surfacing is the topmost layer of carriageway which takes load from traffic directly. It has to provide a smooth non-slippery and stable surface for the vehicles. It should be impervious and should protect base and sub-base from rainwater. It may be provided with bituminous material or with cement mixed with baby jelly.

Shoulders

The width of the carriageway is extended on both the sides by a minimum of 2 – 5 m. It acts as service lane for the broken down vehicle and in case of blocking of carriageway it serves as emergency lane.

The requirements of shoulders are:

- Its colour should be different from that of pavement surface so that they are distinct in vision.
- They should have sufficient load bearing capacity so as to support the loaded trucks in wet weather also.
- Surface of shoulder should be rough when compared to pavement so that drivers are discouraged to use it as regular lane.

Other Components of Roads

Some of the roads will have the following components:

- Traffic separators.
- Kerbs.
- Footpaths.
- Cycle tracks.
- Parking lanes.
- Guard rails.
- Fencing.

Traffic separators are provided to separate the traffic moving in opposite directions. It avoids head on collision between vehicles moving in the opposite direction.

Traffic separators may be in the form of parkway strips or pavement marking whose width vary from 3 to 5 m. If width is to be reduced due to unavoidable situations, 1 in 15 to 1 in 20 transitions are provided.

Kerbs are provided to show the boundary between carriageway and footpaths or shoulder. They provide lateral stability to the base course. There are three classes of kerbs which depend on the height of the kerbs and its function.

Class I kerbs are known as mountable kerbs or low kerbs. Their height with respect to pavement edge varies in the range of 70 to 80 mm. These kerbs allow the vehicles to mount on in case of emergency.

Class II kerbs are known as low speed barriers or urban parking kerbs. Their height above pavement edge varies in the range of 150 mm to 200 mm. They are provided with 25 mm batter to avoid scrapping of tyres, of vehicle.

These kerbs discourage encroachment of slow speed vehicles but at the same time, in case of acute emergency, allows parking of vehicles with some difficulty.

Class III kerbs are known as high speed barriers. Their height varies in the range of 230 mm to 450 mm. They prevent vehicles leaving carriageways. They are usually provided in bridges and hillside roads.

Footpaths are provided for pedestrian to separate them from vehicular traffics. They are generally needed in city roads. The width of footpath is kept 1.3 m or more, based upon the volume of pedestrian traffic.

To encourage pedestrian to use footpath, the surface must be comfortable and smooth. Sometimes parking lanes are provided in cities to streamline vehicle parking.

In urban areas, where cycles are also popular, separate cycle tracks are provided. Generally a minimum width of 2 m cycle tracks are provided.

When the height of fill exceeds 3 m, the guard rails are provided on the edge of shoulders to prevent accidental fall of vehicles down the fill.

In express highways the fencing is provided to avoid animals or other traffic, entering the roads haphazardly. Typical cross-sections of various roads with their components are shown in figures:

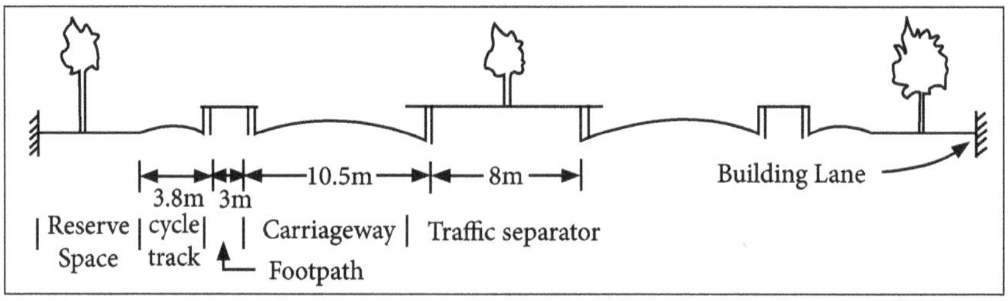

Cross-section of divided highway in urban area.

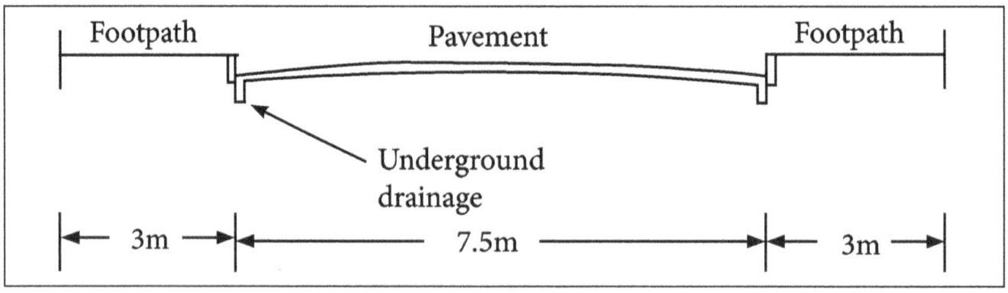

Cross-section of city road in built-up area.

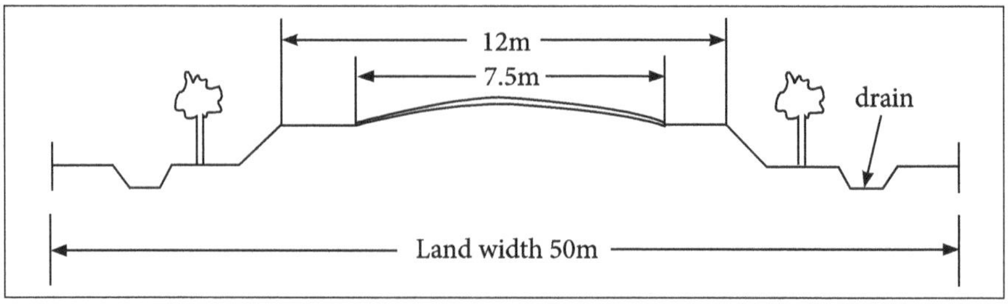

Cross-section of two-lane NH or SH in rural area.

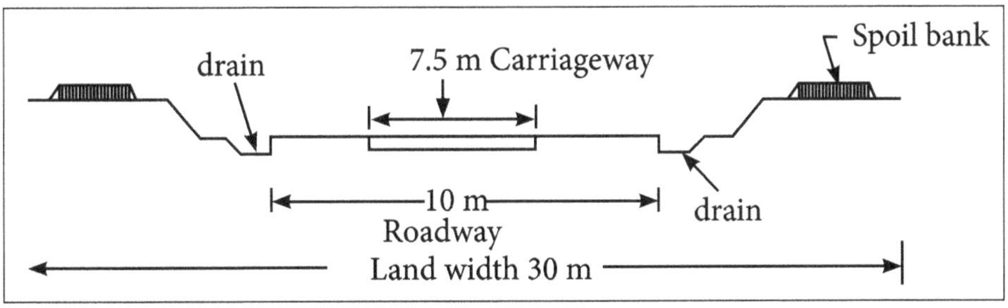

Cross-section of MDR in cutting.

1.3.1 Comparison of Flexible and Rigid Pavements (Advantages and Limitations)

Pavement is the actual travel surface specially made serviceable and durable to withstand the traffic load commuting on it.

Pavement grants friction for the vehicles thus, providing comfort to the driver and transfers the traffic load from the upper surface to the natural soil.

In earlier times before the vehicular traffic became most regular, cobblestone paths were much familiar for on foot traffic load and animal carts.

Pavements are primarily to be used by vehicles and pedestrians. Storm water drainage and environmental conditions are a major concern in the designing of a pavement.

The first of the constructed roads dates back to 4000 BC and it consisted of stone paved streets or timber roads. The roads of the earlier times depended solely on gravel, stone and sand for construction and water was used as a binding agent to level and give a finished look to the surface.

All hard road pavements usually fall into two broad categories namely:

- Flexible pavement.
- Rigid pavement.

Comparison of Flexible and Rigid Pavement

Flexible pavements	Rigid pavements
Deformation in the sub grade is transferred to the upper layers.	Deformation in the subgrade is not transferred to subsequent layers.
Have low flexural strength.	Have high flexural strength.
Design is based on load distributing characteristics of the component layers.	Design is based on flexural strength or slab action.

Load is transferred by grain to grain contact.	No such phenomenon of grain to grain load transfer exists.
Have low completion cost but repairing cost is high.	Have low repairing cost but completion cost is high.
Have low life span when compared to rigid pavements.	Life span is more as compared to flexible pavements.
Surfacing cannot be laid directly on the sub grade but a sub base is needed.	Surfacing can be directly laid on the sub grade.That is why expansion joints are needed.
No thermal stresses are induced as the pavement have the capability to contract and expand freely.	Thermal stresses are more vulnerable to be induced, as the capability to contract and expand is very less in concrete.
Strength of the road is highly dependent on the strength of the sub grade.	Strength of the road is less dependent on the strength of the sub grade.
Rolling of the surfacing is needed.	Rolling of the surfacing in not needed.
Force of friction is less.	Force of friction is high.
Road can be used for traffic within 24 hours.	Road cannot be used until 14 days of curing.
Damaged by oils and certain chemicals.	No damage by oils and greases.

Advantages of Properly Constructed Rigid Pavements

- Low maintenance costs.
- Long life with extreme durability.
- High value as a base, for future resurfacing with asphalt.
- Load distribution over a wide area, decreasing base and sub-base requirements.
- No damage from oils and greases.
- Strong edges.
- Ability to be placed directly on poor soils.

The Limitations of Rigid Pavements

- High initial costs.
- Joints are required for contraction and expansion.
- High repair costs.
- Generally rough riding quality.

Flexible pavements consist of a series of layers, with the highest quality materials at or near the surface.

The strength of a flexible pavement is a result of building up thick layers and thereby, evenly distributing the load over the sub grade, the surface material does not assume the structural strengths as with rigid pavements.

Advantages of Flexible Pavements

- Adaptability to stage construction.
- Availability of low-cost types that can be easily built.
- Resistance to the formation of ice glaze.
- Easy to repair, frost heave and settlement.
- Ability to be easily opened and patched.

Limitations

- Weak edges that may require curbs or edge devices.
- Shorter life span under heavy use.
- Higher maintenance costs.
- Damaged by oils and certain chemicals.

Curbs

- Curbs that are used out of concrete is probably the most common material used for both mountable and barrier curbs. Concrete curbs are relatively easy to construct and durable.
- Granite curbs are not frequently used as asphalt or concrete unless granite is quarried in the area. Granite curbs are more durable than concrete.
- Asphalt is frequently used for curbing, especially where curved parking in lands must be constructed. They are economical and easy to construct.

1.4 Bridges and Culverts: Meaning and Types

A bridge is a structure providing passage over an obstacle without closing the way beneath. The required passage may be for a road, a railway pedestrian or a canal of a pipeline. The obstacle to be crossed may be river, a road, a railway or a valley.

Classification of Bridges

Bridges can be classified into various types depending upon the following factors.

Based on Materials used for Construction

Under this category bridges may be classified as:

- Timber bridges.
- Masonry bridges.
- Steel bridges.
- Reinforced cement concrete bridges.
- Pre-stressed concrete bridges.
- Composite bridges.

Based on Alignment

Under this category, the bridge can be classified as:

- Straight or square bridges.
- Skew bridges.

Straight or square bridges are the bridges which are at right angles to the axis of the river. Skew bridges are not at right angles to the axis of the river.

Based on the Relative Position of Bridge Floor

Under this category, the bridge can be classified as:

- Deck bridge.
- Semi through bridge.
- Through bridge.

Deck bridges are the bridges whose floorings are supported at the top of the super structure.

Through bridges are the bridges whose floorings are supported at the bottom of the super structure. Semi-through bridges are the bridges whose floorings are supported at some intermediate level of the super structure.

Based on Function of Purpose

Under this, the bridge can be classified as:

- Highway Bridge.
- Railway Bridge.
- Foot bridge.
- Viaduct.
- Aqueduct.

Based on Position of High Floor Level

Under this, the bridges may be classified as:

- Submersible bridge.
- Non-submersible bridge.

Submersible bridges are the bridges whose floor levels are below the high flood level. During flood seasons, it allows the water to pass over the bridge submerging the communication route. In economic point of view, these bridges are constructed.

Non-submergible bridges are the bridges whose floor levels are above the high flood level.

Based on Life Span

Under this, the bridges may be classified as:

- Permanent bridges.
- Temporary bridges.

Based on Type of Superstructure

Under this, the bridges may be classified as:

- Arch bridges.
- Truss bridges.
- Portal frame bridges.

- Balanced cantilever bridges.
- Suspension bridges etc.

Based on Span Length

Under this category, the bridges can be classified as:

- Culverts (span less than 6m).
- Minor bridges (span between 6 to 30m).
- Major bridges (span above 30m).
- Long span bridges (span above 120m).

Based on Loading

Road bridges and culverts have been classified according to the loadings they are designed to carry by Indian road congress into:

- Class AA bridges.
- Class A bridges.
- Class B bridges.

Culvert

Culvert is a tunnel structure constructed under roadways or railways to provide cross drainage or to take electrical or other cables from one side to other. The culvert system is totally enclosed by soil or ground.

Materials for Culvert Construction

Culverts are like pipes but very large in size.

They are made of many materials like:

- Concrete.
- Steel.
- Plastic.
- Aluminum.
- High Density Polyethylene.

In most cases concrete culverts are preferred. Concrete culverts may be reinforced or non-reinforced. In some cases culverts are constructed in site called cast in situ culverts.

Precast culverts are also available. By the combination above materials we can also get composite culvert types.

Location of Culverts

The location of culverts should be based on economy and usage. Generally it is recommended that the provision of culverts under roadway or railway are economical.

There is no need to construct separate embankment or anything for providing culverts. The provide culverts should be perpendicular to the roadway. The culverts should be of greater dimensions to allow maximum water level.

The culvert should be located in such a way that flow should be easily done. It is possible by providing required gradient.

Types of Culverts

Following are the types of culverts generally used in construction:

- Pipe culvert (single or multiple).
- Pipe Arch (single or multiple).
- Box culvert (single or multiple).
- Arch culvert.
- Bridge culvert.

Pipe Culvert (Single or Multiple)

Pipe culverts are widely used culverts and rounded in shape. The culverts may be of single in number or multiple. If single pipe culvert is used then larger diameter culvert is installed.

If the width of channel is greater than we will go for multiple pipe culverts. They are suitable for larger flows very well. The diameter of pipe culverts ranges from 1 meter to 6m. These are made of concrete or steel etc.

Pipe Culvert.

Pipe Arch Culvert (Single or Multiple)

Pipe arch culverts means nothing but they looks like half circle shaped culverts. Pipe arch culverts are suitable for larger water flows but the flow should be stable. Because of arch shape fishes or sewage in the drainage easily carried to the outlet without stocking at the inlet or bottom of channel.

This type of culverts can also be provided in multiple numbers based on the requirement. They also enhance beautiful appearance.

Pipe Arch Culvert.

Box Culvert (Single or Multiple)

Box culverts are in rectangular shape and generally constructed by concrete. Reinforcement is also provided in the construction of box culvert. These are used to dispose rain water. So, these are not useful in the dry period.

They can also be used as passages to cross the rail or roadway during dry periods for animals etc. Because of sharp corners these are not suitable for larger velocity. Box culverts can also be provided in multiple numbers.

Box Culvert (Single).

Box Culvert (Multiple).

Arch Culvert

Arch culvert is similar to pipe arch culvert but in this case an artificial floor is provided below the arch. For narrow passages it is widely used. The artificial floor is made of concrete and arch also made of concrete. Steel arch culverts are also available but very expensive.

Arch Culvert.

Bridge Culvert

Bridge culverts are provided on canals or rivers and also used as road bridges for vehicles. For this culvert a foundation is laid under the ground surface. A series of culverts are laid and pavement surface is laid on top this series of culverts.

Generally these are rectangular shaped culverts these can replace the box culverts if artificial floor is not necessary.

Bridge Culvert.

1.4.1 RCC, Steel and Composite Bridges

R.C.C Bridge

R.C.C bridges are constructed using reinforced cement concrete. These are more stable and durable. They can bear heavy loads and are widely using nowadays.

R.C.C Bridge.

Steel Bridge

Steel bridges are constructed using steel bars or trusses or steel cables. These are more durable and bear heavy loads.

Steel Bridge.

Composite Bridges

'Composite' means that the steel structure of a bridge is fixed to the concrete structure of the deck so that the steel and concrete act together, so reducing deflections and increasing strength.

This is done using 'shear connectors' fixed to the steel beams and then embedded in the concrete. Shear connectors can be welded on, perhaps using a 'stud welder' or better still on export work, by fixing nuts and bolts.

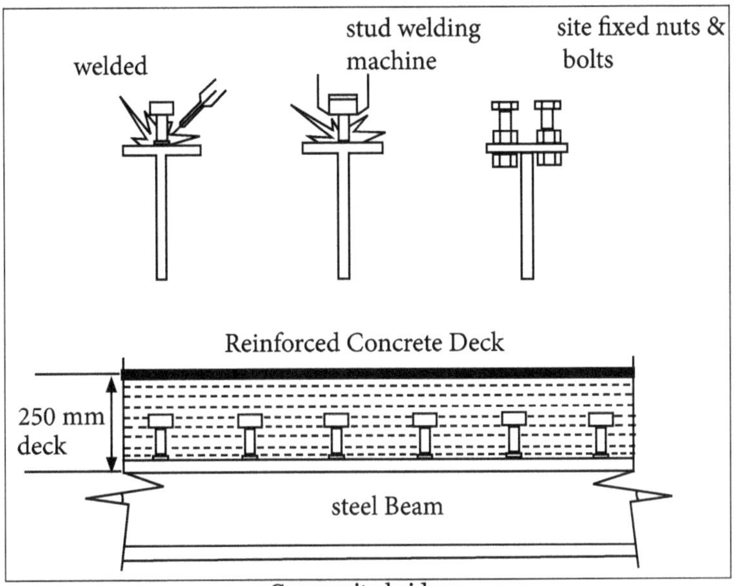

Composite bridges.

Shear connectors, correctly spaced to resist the loads, make the concrete work 'compositely' with the steel.

Usually the steel carries its own weight and that of the wet concrete. But when the concrete is 'cured' and has acquired its full strength, then all future loads (traffic, surfacing, wind, water, pressure, seismic loads) are shared by the steel/concrete composite.

The concrete is good in compression, while the steel is good in tension and compression.

Component Parts of a Bridge

Broadly, a bridge can be divided into two major parts:

- Sub structure.
- Super structure.

Sub Structure

The function of the sub structure is similar to that of foundations, columns and walls of a buildings, because it supports the super structure of the bridges and transmits the load safely to the ground.

The substructure consists of the following:

- Abutments.

- Piers.
- Wing walls.
- Approaches.
- Foundations for the piers and abutments.

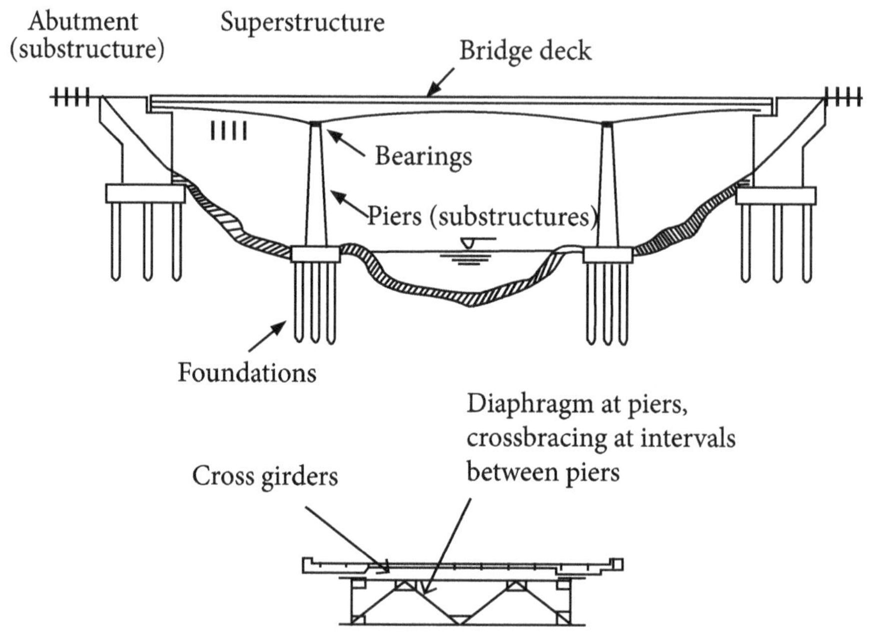

Components of bridge.

Abutments: The end of superstructure of a bridge is called abutments.

Its main functions are:

- To laterally support the earth work of the embankment of the approaches.
- To transmit the load from the bridge superstructure.
- To give final formation level to the bridge.

Bridge abutments can be made of brick masonry, stone masonry, plain concrete or reinforced concrete.

Piers:

Piers are the intermediate supports for the superstructure. Piers transmit the loads from the superstructure of the bridge to the foundations. A pier essentially consists of a column or shaft and a foundation.

They may have different configurations as shown in figure. These piers may be constructed with stone masonry or concrete.

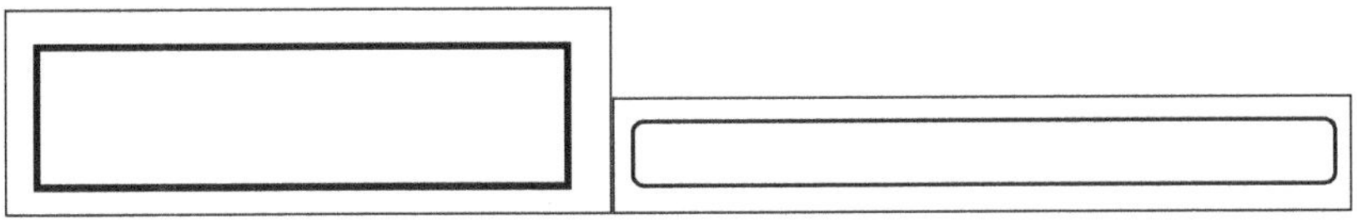

Square nose Rounded nose.

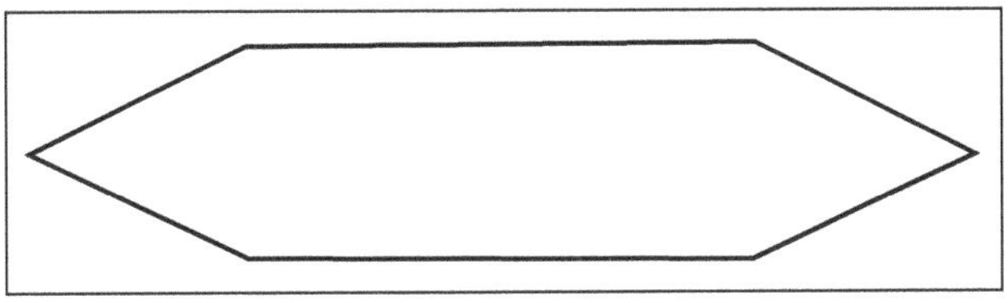

Sharp nose.

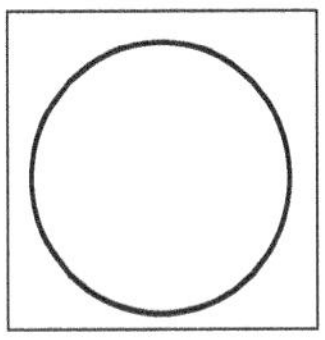

Cylinder.

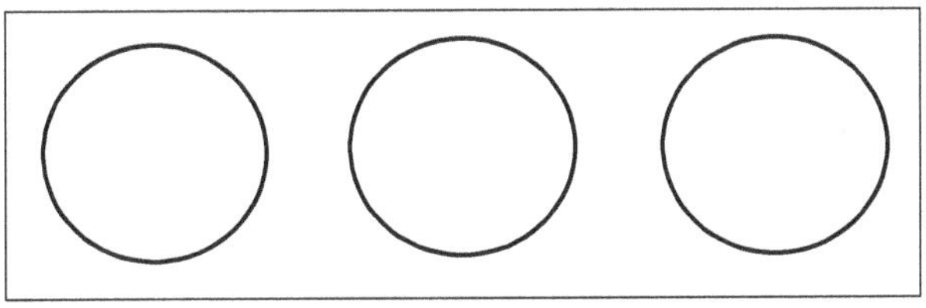

Group of cylinders
Shapes of piers.

Wing Walls:

Wing walls are the walls provided at both ends of the abutments to retain the earth filling of the bridge approaches. They are constructed of the same material as those of the main abutment.

Approaches:

The portion of the road constructed to reach the bridge from their general route or height is known as approach of the bridge. The alignment and the level of the approaches mainly depend on the design and layout of the bridge.

Foundations for the Piers and Abutments:

The foundation of a bridge structure distributes the load from the piers and abutments over the larger area of the sub-soil. It prevents the tilting and over-turning of the piers and abutments and also unequal settlement of the sub-soil.

The different types of functions adopted for bridges are:

- Spread foundation.
- Raft foundation.
- Pile foundation.
- Caisson foundation.
- Well foundation.

Super Structure

The super structure is that part of the bridge over which the traffic moves safely.

It consists of:

- Decking.
- Parapet or hand rails, guard stones etc.
- Bearing.

Decking is provided to allow the road surface to be built in over it. It may consist of a slab, trusses, arches etc.

Parapet or hand rails, guard stones: These are the protective works provided on both sides of the deck along the roadway in order to safe guard the moving vehicles and the passengers on a bridge.

Foot paths are also provided for pedestrians to walk along the bridge. In order to prevent a vehicle from striking the parapet wall of the hand rails, guard stones painted in white are provided at the ends of the road surfaces.

Bearing is the part of the bearing structure provided, to distribute the load coming from the superstructure and also to allow for longitudinal and angular movements.

1.5 Dams: Types

A dam is an impervious barrier construction across a river to store water. The side on which water gets collected is called the upstream side and the other side of the barrier is called the downstream side.

The lake of water which is collected in the upstream side is called as reservoir. This water is then utilized as and when it is needed.

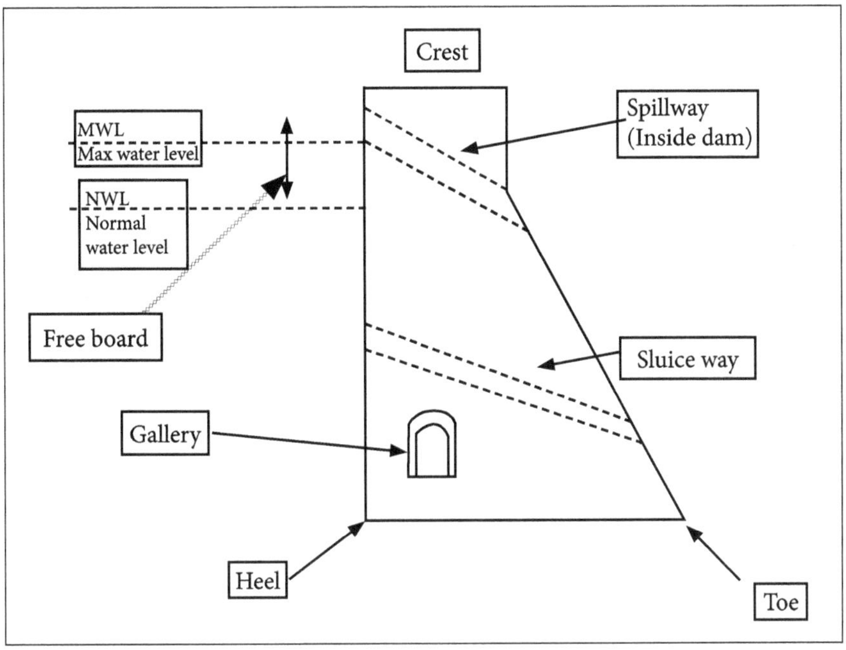

Structure of dam.

Purpose of a Dam

- To store and control the water for irrigation.
- To store and divert the water for domestic uses.
- To supply water for industrial uses.
- To develop hydroelectric power plant to produce electricity.
- To increase water depths for navigation.
- To create storage space for flood control.
- To preserve and cultivate the useful aquatic life.
- For recreational purposes.

Different Types of Dams

Classification According to Material

According to this classification, dam may be classified as follows:

- Rigid dam.
- Non-rigid dam.

Rigid Dams

Rigid dams are those which are constructed of rigid materials such as masonry, concrete, steel or timber.

Rigid dams may be further classified as follows:

- Solid masonry gravity dam.
- Solid concrete gravity dam.
- Arched masonry dam.
- Arched concrete dam.
- Concrete buttress dam.
- Steel dam.
- Timber dam.

Non-rigid Dams

Non-rigid dams are those which are constructed of non-rigid materials such as earth and/or rockfill.

The most common types of non-rigid dams are:

- Earth dam.
- Rockfill dam.

Steel Dams

Steel dams are constructed with a frame work of steel with a thin skin plate as deck slab, on the upstream side.

Steel dams are generally of two types:

- Direct strutted type.
- Cantilever type.

In the direct strutted type, the load on the deck plate is carried directly to the foundation through inclined struts. In the cantilever type, the deck is formed by a cantilever truss i.e. the deck is anchored to the foundation at the upstream toe.

Timber Dams

A timber dam is constructed of framework of timber struts and beams, with timber plan facing to resist water pressure. They are suitable in places where timber can be available in plenty.

Classification According to Structural Behaviour

- Gravity Dam.
- Arch Dam.
- Buttress Dam.
- Embankment Dam.

Type	Material	Section view	Plan(Top view)
Gravity	Concrete, rubble masonry	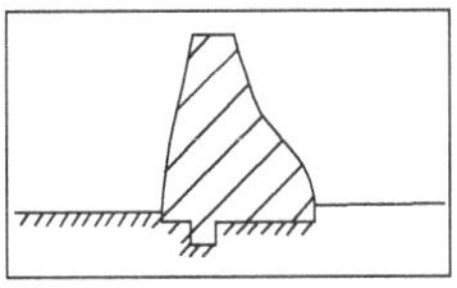	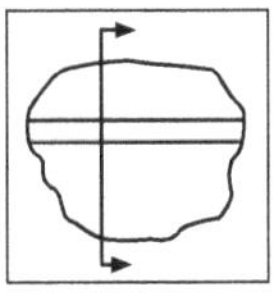
Arch	Concrete,	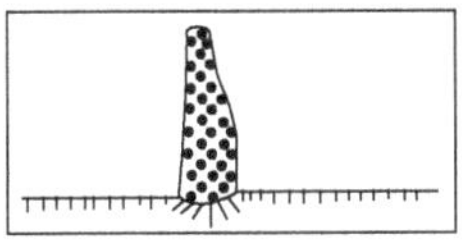	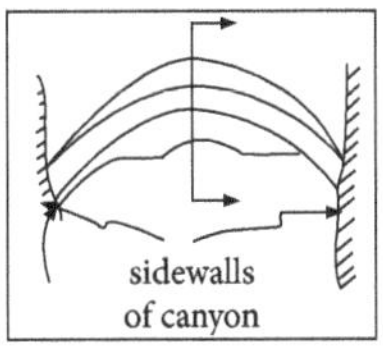
Buttress	Concrete, also timber and steel	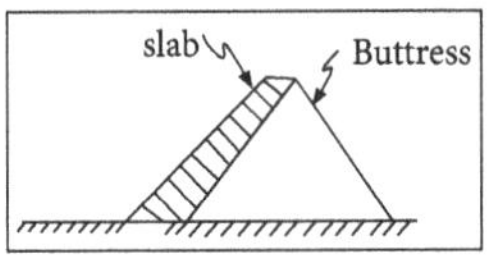	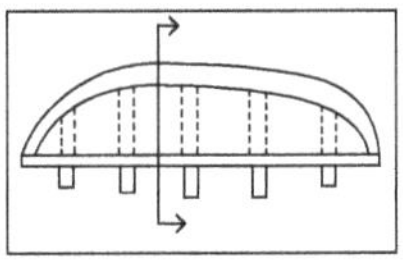
Embankment	Earth or rock	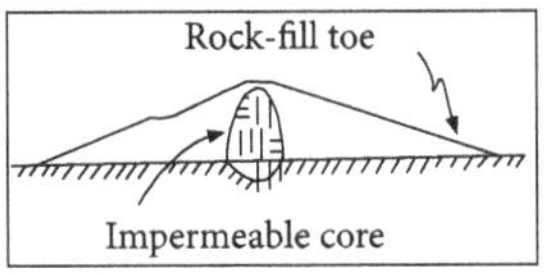	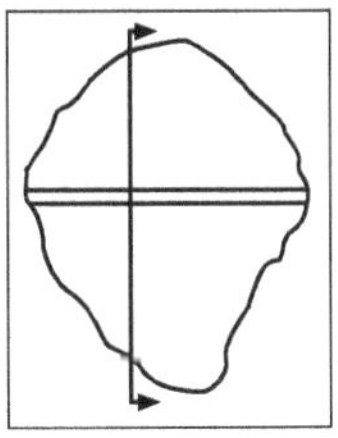

Gravity Dams

A gravity dam is the one in which the external forces such as water pressure, wave pressure, silt pressure, uplift pressure etc. are resisted by the weight of the dam itself. It may be constructed either of masonry or of concrete.

Masonry gravity dams are nowadays constructed of only small heights. All major and important gravity dams are now constructed of concrete only. A gravity dam may be either straight or curved in plan.

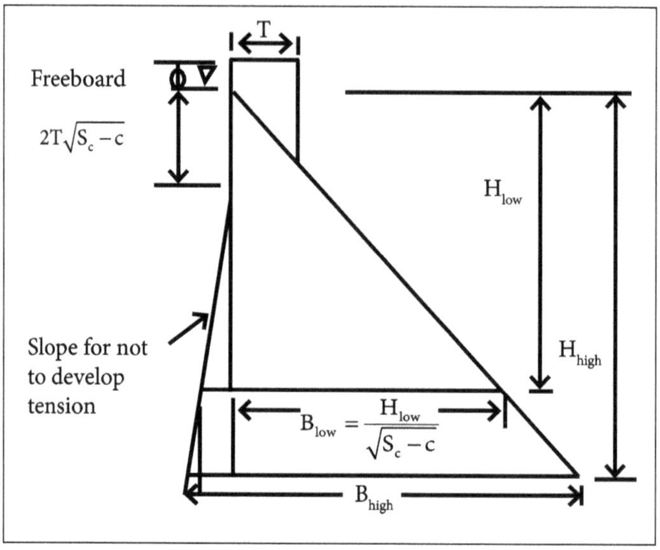

Gravity dam.

Arch Dams

An arch dam is a dam curved in plan and carries a major part of its water load horizontally to the abutments by arch action. The thrust developed by the water load carried by arch action essentially require strong side walls of the canyon to resist the arch forces.

The weight of arch dams is not counted on to assist materially in the resistance of external loads.

Buttress Dams

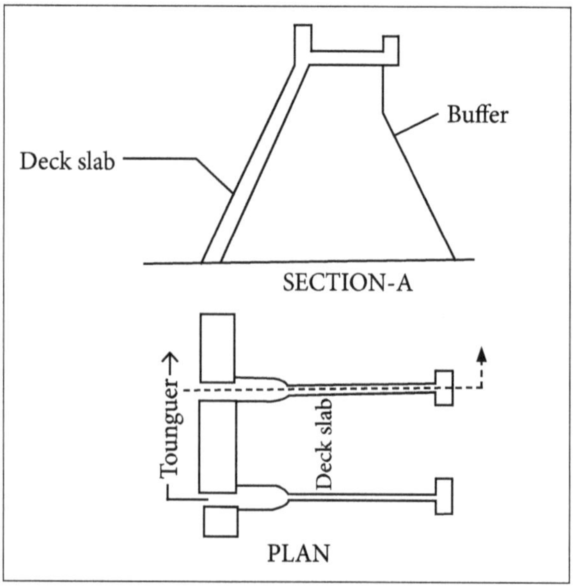

Deck type buttress dam.

A buttress dam consists of a number of buttresses or piers. These piers divide the space into number of spans. Between these piers, panels are constructed of horizontal arches or flat slabs.

When the panels consist of arches, it is known as multiple arches type buttress dam. If the panels consist of flat slab, it is known as deck type buttress dam.

Earth Dams

Earth dams are made of locally available soils and gravels. Therefore, these type of dams are used up to moderate heights only. Their construction involves utilization of materials in the natural state requiring a minimum of processing. Following are the three types of earth dams:

- Homogeneous embankment type.
- Zoned embankment type.
- Diaphragm embankment type.

Homogeneous Embankment Type:

In this type, dam is composed of a single kind of material. But this dam is structurally weak. To check the seepage through the dam a horizontal filter drain or rock toe is provided.

Zoned Embankment Type:

In this type, the dam is made up of more than one material. Usually this dam consists of central impervious core and outer previous shell as shown in figure. A suitable drainage system, in the form of horizontal drain or a rock toe is also provided.

Diaphragm Type Embankment:

In this type, a thin diaphragm of impermeable materials is provided at the centre of the section to check the seepage. The diaphragm may be made of cement masonry, cement concrete or impervious soils.

Rock Fills Dams

In this type, variable sizes of rocks are used to form the embankment. The rock fill dam usually consists of the following four parts:

- Main rock fills at the downstream side.
- Upstream impervious membrane resting on the upstream rock cushion.
- Up stream rock cushion of laid up stone.

- Upstream cut-off to check the sub soil seepage.

Classification According to Functionality

Based on use, dams are classified as follows:

- Storage dam.
- Diversion dam.
- Detention dam.
- Debris dam.
- Coffer dams.

Storage dam is constructed to store water to its upstream side during the periods of excess supply in the river and is used in periods of deficient supply.

Diversion dam supply, raises the water level slightly in the rivers and thus, provides head for carrying or diverting water into ditches, canals or other conveyance systems to the place of use.

Detention dam is constructed to store water during floods and release it gradually at a safe rate when the flood reduces.

Debris dams is constructed to retain debris such as sand, gravel and drift wood flowing in the river with water. The water after passing over a debris dam is relatively clear.

Coffer dams is an enclosure constructed around the construction site to exclude water so that the construction can be done in dry. A cofferdam is thus a temporary dam constructed for facilitating construction.

A coffer dam is usually constructed on the upstream of the main dam to divert water into a diversion tunnel or channel during the construction of the dam.

When the flow in the river during construction of the dam is not much, the site is usually enclosed by the coffer dam and pumped dry. Sometimes a coffer dam on the downstream of the dam is also required.

Classification According to Hydraulic Design

According to hydraulic design, dams may be classified as follows:

- Non - overflow dam.
- Overflow dam.

Non-overflow Dam

A non-overflow dam is the one in which the top of the dam is kept at a higher elevation than the maximum expected high flood level.

Over Flow Dam

An overflow dam is the one which is designed to carry surplus discharge over its crest. Usually, in a river valley project, the two types of dams are combined. The main dam is kept as a non-overflow dam and some portion of dam is kept as overflow dam, at some suitable location along the main dam.

Selection of Site

Factors Governing Selection of Site for Dam

- The river cross-section at the dam site should preferably have a narrow gorge to reduce the length of the dam. However, the gorge should open out u/s to provide large basin for a reservoir.
- Suitable foundations should be available at the site selected for a particular type of dam. For gravity dams, sound rock is essential. For earth dams, any type of foundations is suitable with proper treatment.
- A suitable site for the spillway should be available in the near vicinity. If the spillway is to be combined with the dam, the width of the gorge should be such as to accommodate both.
- The general bed level at dam site should preferably be greater than that of the river basin. This will reduce the height of the dam and will facilitate the drainage.
- The reservoir basin should be reasonably water tight. The stored water should not escape out through its side walls and bed.
- To establish site for labour colonies, a healthy environment should be available in the near vicinity.
- Materials required for the construction should be easily available, either locally or in the near vicinity, so that the cost of transporting them is as low as possible.
- The dam site should be easily accessible so that it can be connected economically to important cities and towns.
- The value of land and property submerged by the proposed site should be as low as possible.

1.6 Engineering Mechanics

Engineering mechanics is a vital analytical tool that supports an engineer to optimize a design, making one that it is strong and rigid enough to do the job, but not overly heavy and expensive.

Virtually any physical device in the "real world' is acted on by forces, which can include gravitational, pressure, electrostatic, magnetic, centrifugal and impact forces.

Engineering mechanics studies the effects that forces have on materiels allowing engineers to theories how a design will react before it is built. Engineers may then optimize designs on paper, without having to build and test multiple versions of a product.

By using mechanics, better products can be designed and built, faster and more cost effectively. Mechanics of materials considers, how forces act upon actual bodies like bending, twisting or breaking them.

Engineers are using different design criteria to create physical products, including design for strength, factor of safety and design for stiffness.

Depending upon the body to which the mechanics is applied, the engineering mechanics is classified as:

- Mechanics of Solids.
- Mechanics of Fluids.

The solid mechanics is further classified as mechanics of rigid bodies and mechanics of deformable bodies. The body which will not deform or the body in which deformation can be neglected in the analysis, are called as Rigid Bodies.

The mechanics of the rigid bodies dealing with the bodies at rest is termed as Statics and that dealing with bodies in motion is called Dynamics.

The dynamics dealing with the problems without referring to the forces causing the motion of the body is termed as Kinematics and if it deals with the forces causing motion also, is called Kinetics.

If the internal stresses developed in a body are to be studied, the deformation of the body should be considered. This field of mechanics is called Mechanics of Deformable Bodies/Strength of Materials/Solid Mechanics. This field may be further divided into Theory of Elasticity and Theory of Plasticity.

Liquid and gases deform continuously with application of very small shear forces. Such materials are called Fluids. The mechanics dealing with behaviour of such materials is called Fluid Mechanics.

Mechanics of ideal fluids, mechanics of viscous fluids and mechanics of incompressible fluids are further classification of Fluid mechanics. The classification of mechanics is summarized below in flowchart.

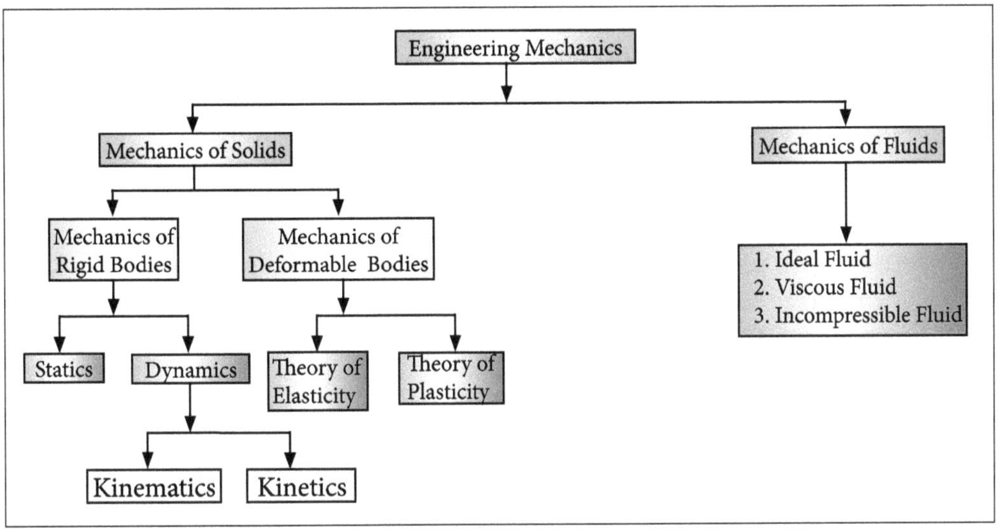

Mechanics

This deals with the study and prediction of the state of rest or motion of particles and bodies under the action of forces.

Kinematics is purely a geometrical description of motion and deformation of material bodies.

Kinetics is, addressing the forces as external actions and the stresses as internal reactions.

Balance equations is used for conservation of mass, momentum and energy and material dependent laws.

The following are the fundamentals terms to study mechanics, which should be understood clearly:

Mass

It is a measure of quality of matter contained by the body. It will not change unless the body is damaged and part of it is physically separated. When a body is taken out in a spacecraft, the mass will not change however its weight may change due to change in gravitational force.

Even the body may become weightless once gravitational force vanishes but the mass remains the same.

Time

Time is the measure of succession of events. The consecutive event selected is the rotation of earth about its own axis and this is called a day.

To have convenient units for numerous activities, a day is split into 24 hours, an hour into 60 minutes and a minute into 60 seconds. Clocks are the instruments developed to measure time. Unit of time is second.

Space

The geometric region in which study of body is involved is called space. A point in the space may be referred with respect to a predetermined point by a set of linear and angular measurements. The reference point is called the origin and set of measurements as 'coordinates'.

If coordinates involve only in mutually perpendicular directions, they are known as Cartesian coordinates. If the coordinates involve angle and distances, it is termed as polar coordinate system.

Length

It is a concept to measure linear distances. The diameter of a cylinder may be 300 mm, the height of a building may be 15 m. actually meter is the unit of length. However, depending upon the sizes involved micro, milli or kilo meter units are used for measurement.

A meter is defined as length of the standard bar of platinum-iridium kept at the International Bureau of Weights and Measures.

Displacement

Displacement is defined as the distance moved by a body/particle in the specified direction. Referring to Figure, if a body moves from position A to position B in the x-y plane shown, its displacement in x-direction is AB′ and its displacement in y-direction is B′B.

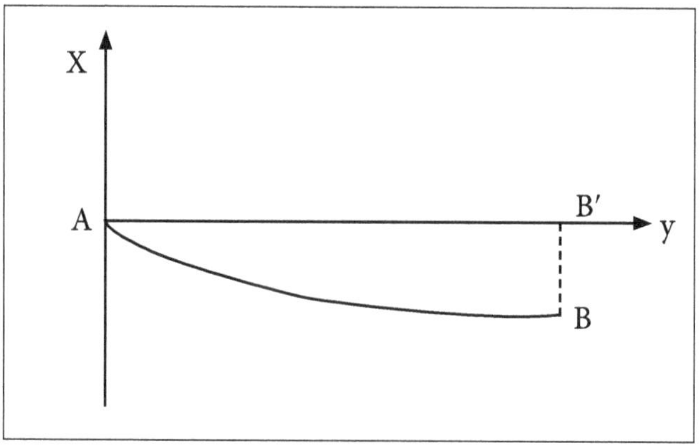

Velocity: The rate of change of displacement with respect to time is defined as velocity.

Acceleration: Acceleration is the rate of change of velocity with respect to time.

Thus,

$$a = \frac{d\upsilon}{dt}, \text{where } \upsilon \text{ is velocity}$$

Momentum: The product of mass and velocity is called momentum.

Thus,

$$\text{Momentum} = \text{Mass} \times \text{Velocity}$$

1.6.1 Basic Idealizations: Particle, Continuum and Rigid Body

In the study of mechanics and solution of problems, a number of ideal conditions are assumed to simplify the formulations.

Actually, in the absence of these idealized concepts, the solution to certain mechanics problems becomes difficult. In addition, the use of these concepts does not introduce appreciable errors in the description and prediction of the state of the system.

Following are the basic idealization adopted in engineering mechanics:

- Particle.
- Continuum.
- Rigid body.
- Point force.

Particle

A particle may be defined as an idealized body which may have finite or negligible mass and whose size and shape can be ignored without introducing appreciable error. Sometimes the particle may be viewed as a mass point or a corpuscle.

In statics, if the lines of action of all the forces acting on a body are concurrent, then the body can be treated as a particle, situated at the point of concurrency. In the case of kinematics, if only translation of the body is considered, then body can be idealized as a particle.

On the other hand, in kinetics, if the unbalanced force exerts an exclusive translational effect, then the body can be assumed as a particle.

Examples are: a bomb released from an aeroplane, motion of earth in its orbit around the sun.

Continuum

An idealized body in which the matter is assumed to be distributed continuously without any void or pore is called continuum. In this, the body is considered to be continuum and the molecular structure of the matter within the body is disregarded.

This theory is particularly useful in the demonstration of the matter at macroscopic level. Especially, in the study of mechanics of fluids, this concept plays a major role.

Rigid Body

A rigid body is an idealized body composed of large number of particles, which remains at a fixed distance from each other. In practice, no body behaves as absolutely rigid under the action of external forces.

However, when the deformations of a body are negligibly small, the body can be regarded as rigid. Figure below illustrates the idealization of a ball indicating the concept of continuum, rigid body and particle.

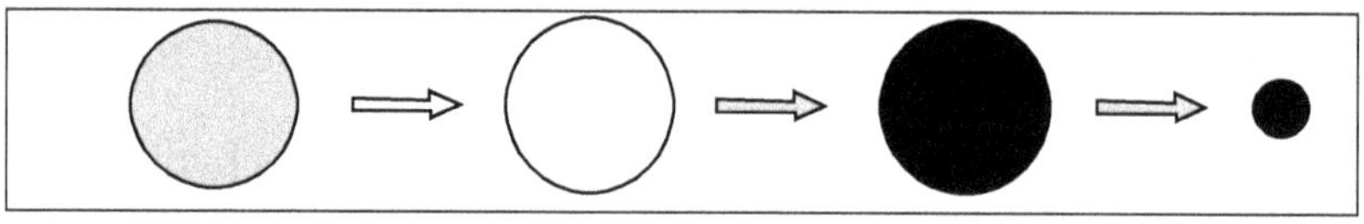

Collection of Molecules Continuum Rigid Body Particle Idealization of a ball.

Point Force

A point force or a concentrated force is an idealized force assumed to be acting at a point of a continuum. The weight of a body, for example, is considered to be single concentrated force acting at the centre of gravity (CG) of the body, in spite of it being distributed throughout the body.

Further, the contact force exerted between two bodies in contact is actually distributed over the contact area. If the contact area is relatively small, the contact force between the two bodies may be regarded as a point force. This idealization facilitates in the analysis and solution of problems.

1.6.2 Newton's Laws of Force and its Characteristics

Newton's Law

For a couple centuries before Einstein, Newton's Laws were the basic principles of the Physics. These laws are still valid and they are the basis for much engineering analysis today. Informal explanations of Newton's Three Laws are given.

Newton's First Law

An object at rest tends to stay at rest and an object in motion tends to stay in motion with the same speed and in the same direction unless acted upon by an unbalanced external force.

Inertia is the property of matter that resists changes in motion. If a mass is not moving, then it will stay that way until an unbalanced external force starts to move it. If a mass is in motion, it will stay in motion with the same speed and direction until an unbalanced external force changes its motion characteristics.

If the puck is placed down on the ice, it will stay motionless until someone hits it with a stick or skate because of its inertia. Also due to inertia, when slapped, the puck will tend to move in a straight line with constant speed until an external force changes its motion.

As a second example of Newton's First Law, consider a car accelerating from a stoplight. As the car accelerates from zero motion, our body tends to push back into the seat due to its inertia (trying to remain at rest).

Also, as the car is braked from a high speed back to stopping, our body is flung forward due to its inertia in motion. Hopefully we have our seatbelt on or else Newton's First Law could have bad consequences.

Newton's Second Law

The acceleration a of an object as produced by a net force F is directly proportional to the magnitude of the net force, in the same direction as the net force and inversely proportional to the mass m of the object: F = ma.

A resultant external force F acting on a body will accelerate that body in the direction of F, with acceleration a = F/m. Acceleration is the second time rate of change of position, also the first time rate of change of velocity; acceleration is to velocity what velocity is to position. Newton's original statement of the Second Law was that the resultant external force F is equal to the time rate of change of momentum (mv, mass time's velocity),

$$F = \frac{d}{dt(mv)}$$

If the mass is constant, this relationship becomes the familiar form of Newton's Second Law,

$$F = m\frac{dv}{dt} = ma$$

Before Newton developed his Second Law, the prevailing belief was that force was proportional to velocity: F = mv. This appeared to be true for the motion of horse-drawn carts, since friction dominates this problem.

Newton revolutionized engineering mechanics; his laws were unchallenged until Einstein's Relativity work. Newton's Laws are still the basis for most engineering dynamics today.

Newton's Third Law

For every action, there is an equal and opposite reaction.

A force cannot be applied to an object unless something resists the reaction of that force. In order to walk across the floor, we must push back on the floor with our foot then, according to Newton's Third Law, the floor pushes forward on our foot, which propels us forward. This, of course, requires friction to work.

If a free-floating astronaut were to throw a baseball, there is nothing to resist the throwing force, so as the baseball accelerates in the direction of throwing, astronaut would accelerate them backwards, with a force equal and opposite to the throwing force.

The astronaut would accelerate at a much smaller level since her/his mass is much greater than the baseball's mass. The recoil of the gun during firing is another example of Newton's Third Law.

1.6.3 Types of Forces: Gravity, Lateral and its Distribution on Surfaces

Force

A force may be defined as the action of one body on another, which may be exerted either by contact or at a certain distance.

It is a derived unit. It is a force that imparts an acceleration of 1 m/s^2 on a body of mass one Kg. 1N = 1 Kg m/s^2 = It is an agency which changes or tends to change the state of rest or motion of a body.

Force has the capacity to impart motion to a particle. Force can produce push, pull or twist. It is a vector quantity, hence to define force its point of application, its direction and its magnitude has to be specified.

For simplicity sake, all forces between objects can be placed into two broad categories.

Gravity

The force of gravity is the force with which the earth, moon or other massively large object attracts another object towards itself. By definition, this is the weight of the object.

All objects upon earth experience a force of gravity that is directed "downward" towards the center of the earth. The force of gravity on earth is always equal to the weight of the object as found by the equation:

$$F_{grav} = m * g$$

Where,

g = 9.8 N/kg (on Earth),

And,

m = mass (in kg).

Contact Forces

This is a type of force in which the two interacting objects are physically contacting each other. Examples of contact forces include frictional forces, tensional forces, normal forces, air resistance forces and applied forces.

Action-At-a-distance Forces

This is a type of force in which the two interacting objects are not in physical contact with each other, yet are able to exert a push or pull despite a physical separation.

Examples:

Gravitational forces (E.g., when our feet leave the earth and we are no longer in contact with the earth, there is a gravitational pull between us and the Earth, the sun and planets exert a gravitational pull on each other despite their large spatial separation).

Electric forces (E.g., the protons in the nucleus of an atom and the electrons outside the nucleus experience an electrical pull towards each other despite their small spatial separation).

Magnetic forces (E.g., 2 magnets can exert a magnetic pull on each other even when separated by a distance of a few cm).

Apart from this, the force is also sub-divided as internal and external force. Internal forces are those that hold together the particles that form the rigid body. If the rigid body has several parts, the forces holding the component parts together are also known as internal force.

External forces represents the action of other bodies on the rigid body which is under consideration. They will either cause it to move or assure that it remains at rest.

1.6.4 Classification of Force Systems

Based upon their relative positions, points of applications and lines of actions, the different force systems can be classified as follows:

- Collinear forces.
- Coplanar parallel forces.
- Coplanar concurrent forces.
- Coplanar non-concurrent forces.
- Non- coplanar parallel forces.
- Non- coplanar concurrent forces.
- Non- coplanar non-concurrent forces.

Collinear Forces

It is a force system, in which all the forces have the same line of action.

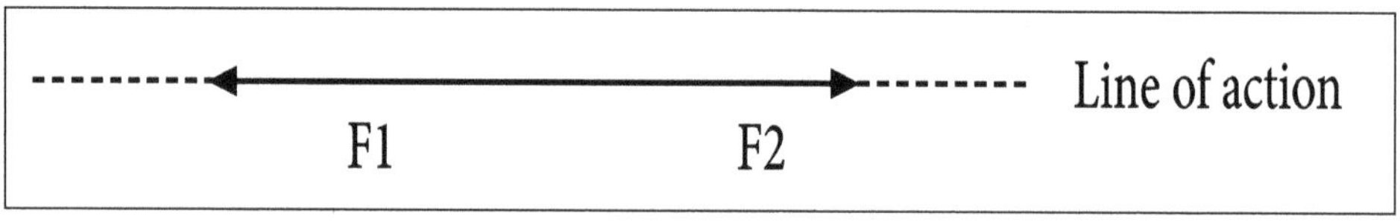

Eg: Forces on a rope in tug of war.

Coplanar Parallel Forces

It is a force system, in which all the forces are lying in the same plane and have parallel lines of action.

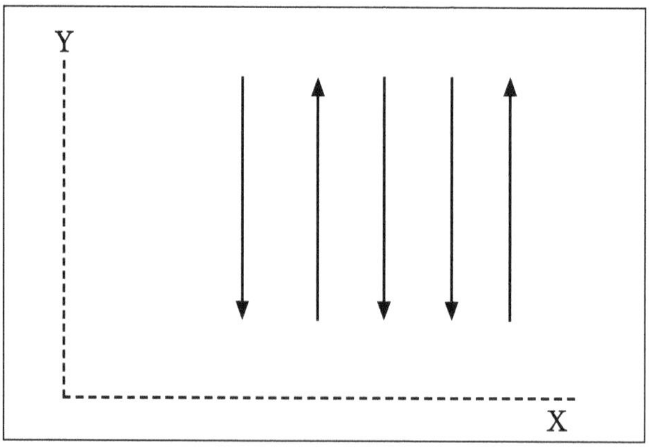

Eg: The forces or loads and the support reactions in case of beams.

Coplanar Concurrent Forces

It is a force system, in which all the forces are lying in the same plane and lines of action meet at a single point.

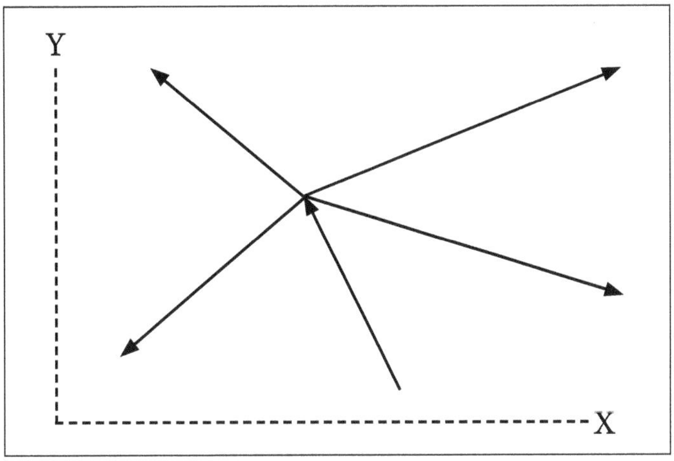

Eg: The forces in the rope and pulley arrangement.

Coplanar Non-concurrent Forces

It is a force system, in which all the forces are lying in the same plane but lines of action do not meet at a single point.

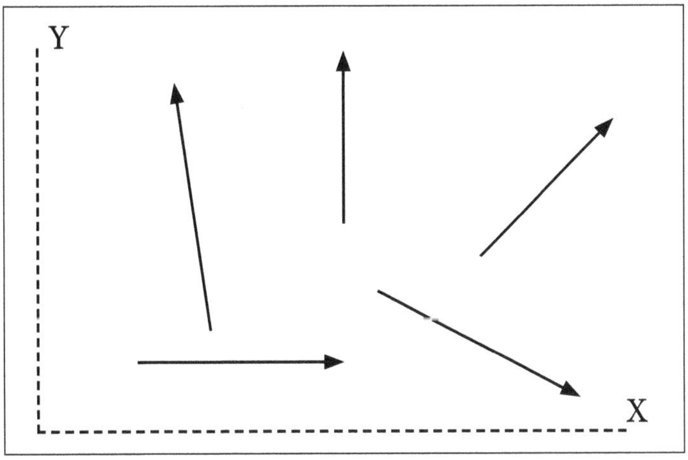

Eg: Forces on a ladder and reactions from floor and wall, when a ladder rests on a floor and leans against a wall.

Non- coplanar Parallel Forces

It is a force system, in which all the forces are lying in the different planes and still have parallel lines of action.

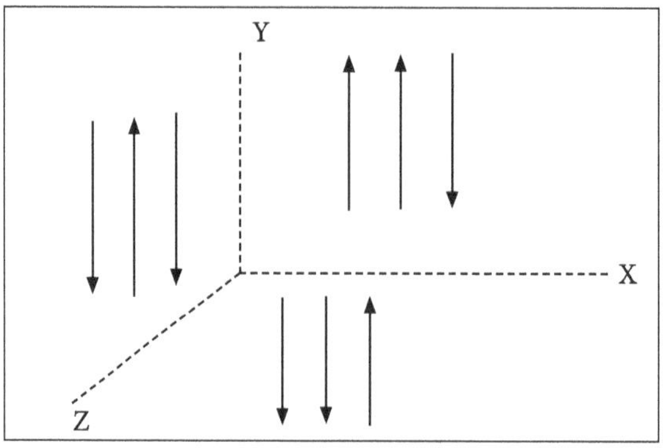

Eg: The forces acting and the reactions at the points of contact of bench with floor in a classroom.

Non-coplanar Concurrent Forces

It is a force system, in which all the forces are lying in the different planes and still have common point of action.

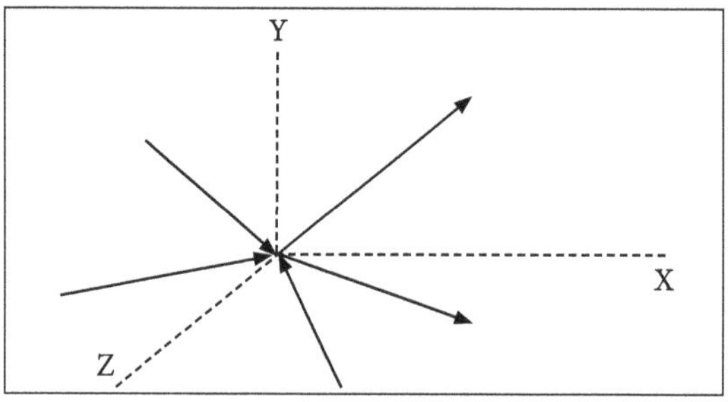

Eg: The forces acting on a tripod when a camera is mounted on a tripod stand.

Non-coplanar Non-concurrent Forces

It is a force system, in which all the forces are lying in the different planes and also do not meet at a single point.

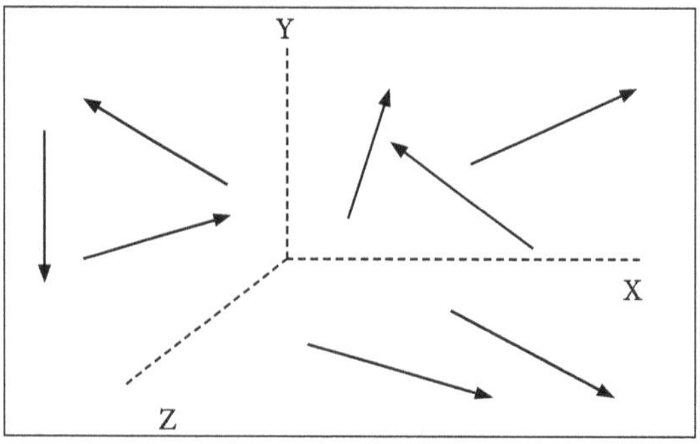

Eg: Forces acting on a building frame.

1.6.5 Principle of Physical Independence and Superposition and Transmissibility of Forces

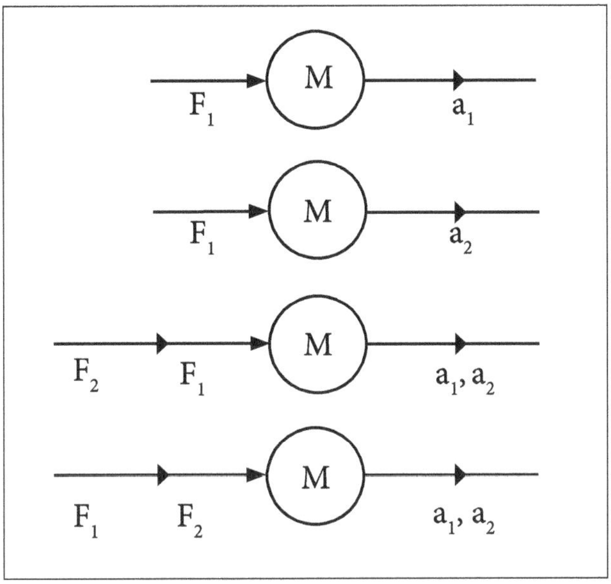

Action of forces on the bodies are independent, in other words the action of forces on a body is not influenced by the action of any other force on the body.

Principle of Superposition of Forces

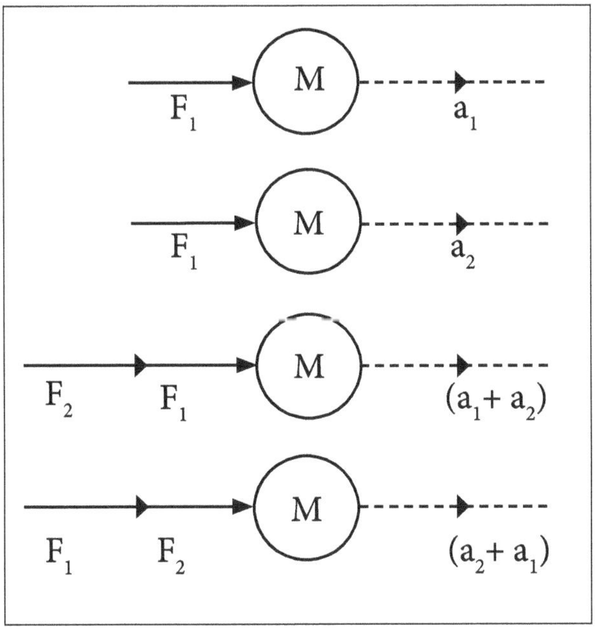

Net effect of forces applied in any sequence on a body is given by the algebraic summation of effect of individual forces on the body.

Principle of Transmissibility of Forces

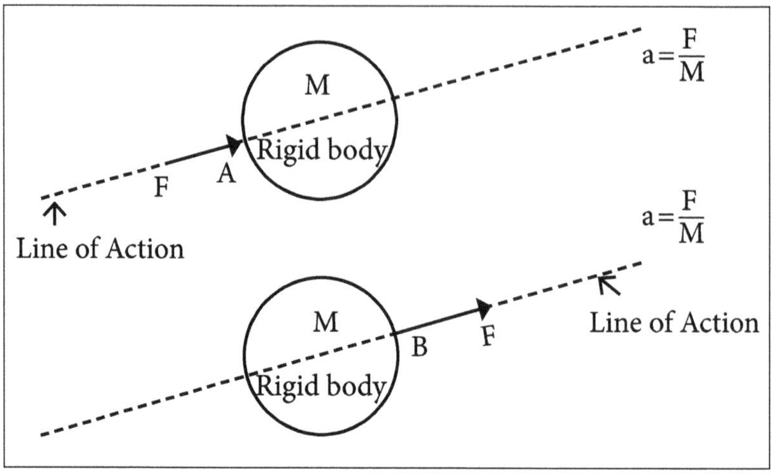

The point of application of a force on a rigid body can be changed along the same line of action maintaining the same direction and magnitude without affecting the effect of the force on that body.

Limitation of Principle of Transmissibility:

It can be used only for rigid bodies and cannot be used for deformable bodies.

1.6.6 Introduction to SI Units

The SI system is the most modern form of the metric system and it can be considered as rationalized MKS system. This system has been adopted by the ISO as the abbreviated name for the system of units in all languages.

The SI system is simply an expanded form of the RMKSA system. RMKSA stands for Rationalized Meter, Kilogram, Second, Ampere. In addition to these, Degree Kelvin (unit of absolute temperature), Mole (amount of substance) and Candela (unit of luminous intensity) are included in the SI system of units.

This latest system of units was introduced mainly for wide application in all branches of engineering and science, but it is also applicable in other fields.

Prefixes and Symbols of Multiplying Factors in SI

Multiplying factor	Prefix	Symbol
10^{12}	tera	T
10^{9}	giga	G
10^{6}	mega	M
10^{3}	kilo	k

10^{0}	-	-
10^{-3}	milli	m
10^{-6}	micro	µ
10^{-9}	nano	n
10^{-12}	pico	p
10^{-15}	femto	f
10^{-18}	atto	a

Definitions of the SI Base Units

Unit of length	Meter	The meter is the length of the path travelled by light in vacuum, during a time interval of 1/299 792 458 of a second.
Unit of amount of substance	Mole	1. The mole is the amount of substance of a system which contains as many elementary entities as there are atoms in 0.012 kg of carbon 12; its symbol is "mol." 2. When mole is used, the elementary entities should be specified and may be atoms, electrons, molecules, ions and other particles or specified groups of such particles.
Unit of electric current	Ampere	The ampere is defined as that constant current which, if maintained in 2 straight parallel conductors of infinite length and of negligible circular cross-section, placed one meter apart in vacuum, would produce a force equal to 2 x 10-7 newton per meter of length between these conductors.
Unit of time	Second	The second is the duration of 9 192 631 770 periods of the radiation corresponding to the transition between the 2 hyperfine levels of the ground state of the cesium 133 atom.
Unit of thermodynamic temperature	Kelvin	The kelvin, unit of thermodynamic temperature, it is the fraction of 1/273.16 of the thermodynamic temperature of the triple point of water.
Unit of luminousintensity	Candela	The candela is the unit of luminous intensity, in a specified direction, of a source that produce monochromatic radiation of frequency 540 x 1012 hertz and that has a radiant intensity in that direction of 1/683 watt per steradian.
Unit of mass	Kilogram	The unit of mass is kilogram; it is equal to the mass of the international prototype of the kilogram.

1.7 Couple, Moment of a Couple, Characteristics of a Couple and Moment of a Force

Couple

When two parallel forces that have the same magnitude but in opposite direction is known as couple. The couple is separated by perpendicular distance. As matter of fact a couple is unable to produce any straight-line motion but it produces rotation in the body on which it acts.

So couple can be defined as unlike parallel forces of same magnitude but opposite direction which produce rotation about a specific direction and whose resultant is zero.

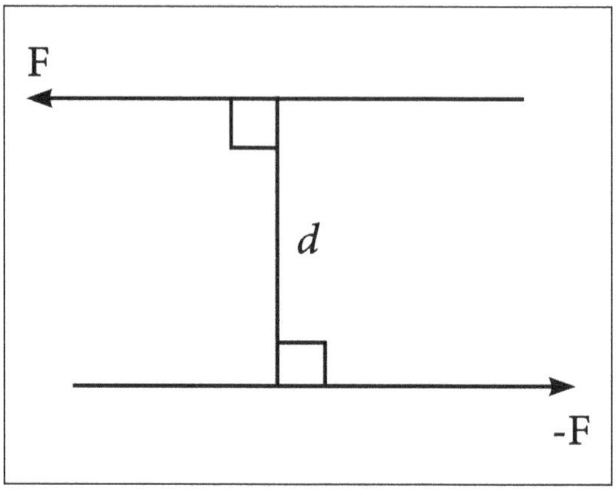

A couple has a tendency to produce a moment about the body or to rotate a body. As such, the moment due to a couple is denoted as M. Let us consider a point O about which a couple acts. Let S be the distance separating the couple. Let d_1 & d_2 be the perpendicular distance of the lines of action of the forces from the point O.

Thus, the magnitude of the moment due to the couple is given as,

$$M_O = (F \times d_1) + (F \times d_2)$$

$$M_O = F + d$$

i.e The magnitude of a moment due to a couple is defined as the product of the force constituting the couple & the distance separating the couple. Hence the units for magnitude of a couple can be N m, kN m, N mm, etc.

The moment of a force is a vector which is the product of distance and force. Hence in case of moment* of a force the cross-product of distance and force would be taken. Consider the figure below.

Let,

F = Force vector $(F_x i + F_y j + F_z k)$

r = Distance (or position) vector with respect to O.

$= x_i + y_j + zk$

M = Moment of force about point O.

A quantity which is the product of two vectors and the quantity is also a vector, then cross product of the two vectors will be taken. But if the quantity is scalar, then dot product is taken.

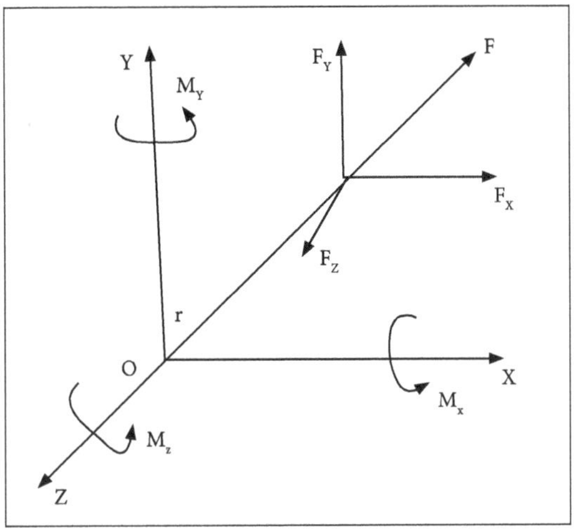

Then,

$$M = r \times F$$

or,

$$M = r \times F = \begin{vmatrix} i & j & k \\ x & y & z \\ F_x & F_y & F_z \end{vmatrix}$$

$$\therefore M = (yF_z - zF_y)i + (zF_x - xF_z)j + (xF_y - yF_x)k$$

The moment of the given force about x, y and z - axis are equal to,

$$M_x = yF_z - zF_y,\ M_y = zF_x - xF_z,\ M_z = xF_y - yF_x$$

Where,

M_x = Moment of F about x-axis,

M_y = Moment of F about y-axis and

M_z = Moment of F about z-axis.

Also,

M_x, M_y and M_z are known as scalar components of moment.

Vectorial Representation of Couples

The moment produced by two equal, opposite parallel forces is known as couple. Figure shows two equal opposite and parallel forces acting at points A and B. Let r_A and r_B be the position vectors of A and B with respect to O. The vector which joins B to A is represented by r.

The moment of two forces about point O is given by,

$$M_o = r_A \times F - r_B \times F = (r_A - r_B) \times F = r \times F (r_A - r_B = r)$$

The above equation shows that moment vector is independent of moment centre O.

$$M = r \times F$$

This moment is known as couple.

The effect of couple is to produce pure rotation about an axis normal to the plane of force which constitute couple.

Application of Couple

- To open or close the valves or bottle head, tap etc.
- To wind up a clock.
- To move the paddles of a bicycle.
- Turning a key in lock for open and closing.

Moment of a Couple

The magnitude of the moment of the couple is determined by using the principle of superposition.

That is, the moment of the couple is equal to the sum of the moment of the two forces of the couple about any point.

As seen in figure (a), the moment of couple about O_1 is given by,

$$M_{O1} = +F(d_1) - F(a + d_1) = -F \times a$$

Similarly, the moment of the couple about point O_2 is,

$$M_{O2} = -F(a - d_2) - F(d_2) = -F \times a$$

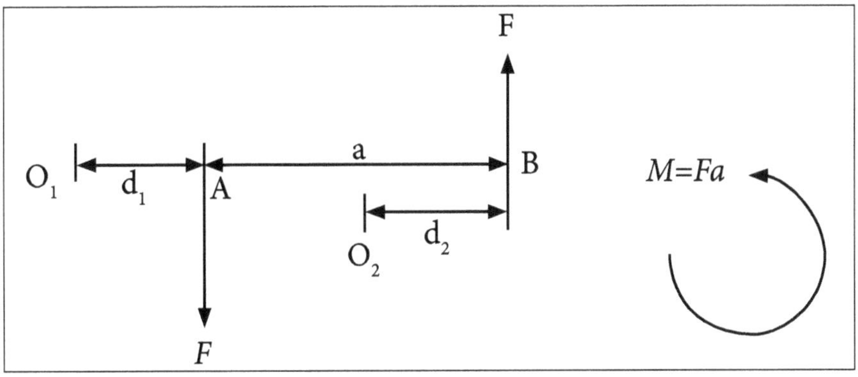

(a) (b)

Moment of a couple.

It is clear that the moment of a couple about any point is always constant. Interestingly, couple can also be diagrammatically shown by a rotational arrow as shown in figure (b) indicating the magnitude of the moment of a couple. M = Fa.

Characteristics of Couple

The main characteristics of couple are:

- The algebraic sum of the forces, having the couple, is zero.
- The algebraic sum of moment of the forces, constituting couple, about any point is the same and equal to the moment of couple itself.
- Any number of coplanar couples can be reduced to single couple, whose magnitude will be equal to algebraic sum of moments of all the couples.
- A couple can't be balanced by a single force, but can be balanced only by a couple, however of opposite sense.

Moment of Force

It is defined as the rotational effect caused by a force on a body. Mathematically, moment is defined as the product of the magnitude of the force and the perpendicular distance of the point from the line of action of the force to that point where the object will turn.

The Moment of a force is a measure of its tendency to cause a body to rotate about a specific point or axis. This is different from the tendency for a body to move or translate, in the direction of the force.

In order for a moment to develop, the force must act upon the body in such a manner that the body would begin to twist. This occurs every time a force is applied so that it does not pass through the centroid of the body.

A moment is due to a force not having an equal and opposite force directly along its line of action.

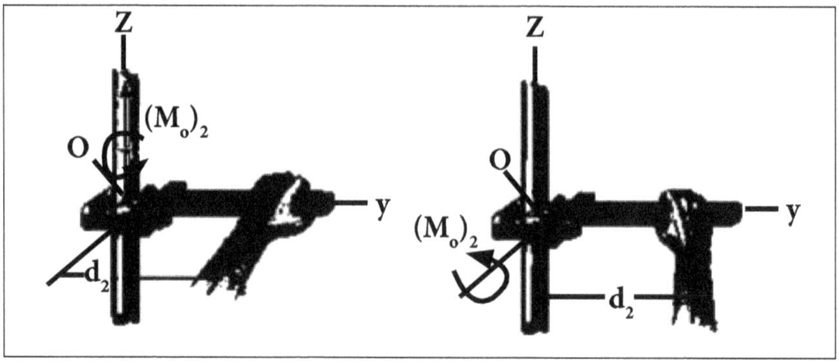

(a) (b)

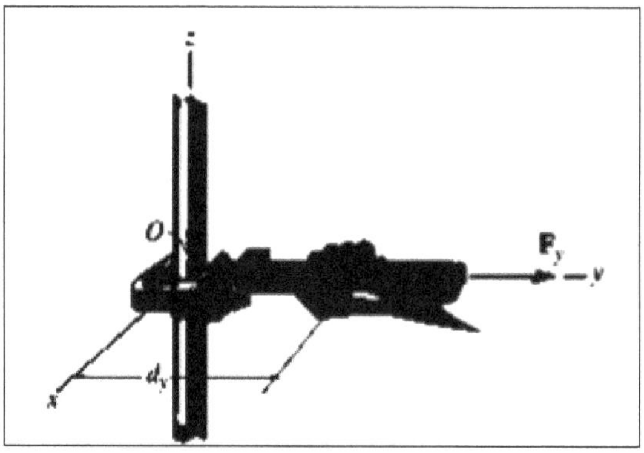

(c)

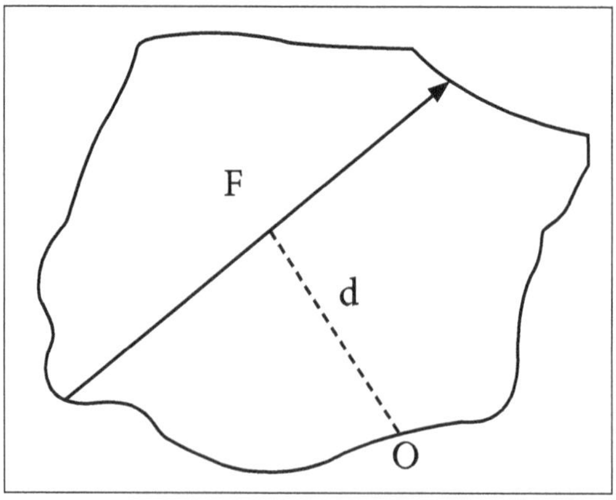

Let "O" be a point or particle in the plane. Let "d" be the perpendicular distance of the line of action of the force from the point "O". Let "F" be a force acting in a plane.

Thus, the moment of the force about the point "O" is given as,

$$M_O = F \times d$$

Rotational effect or moment of a force is a physical quantity dependent on the units for force and distance. Hence the units for moment can be "KNm" or "Nm" or "N mm" etc. The moment produced by a force about different points in a plane is different. This can be understood from the following figures:

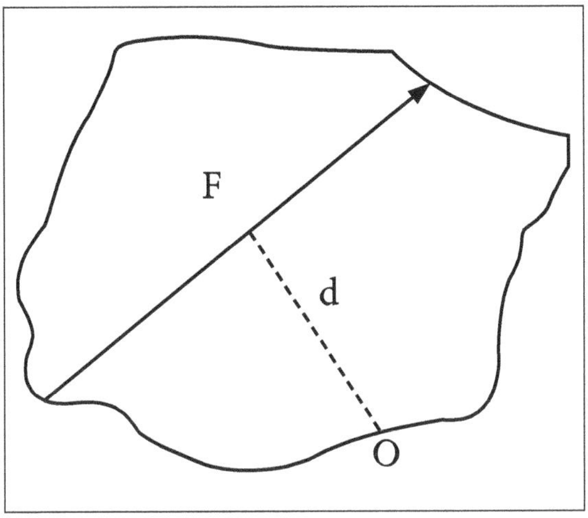

Let "F" be a force in a plane and O_1, O_2 and O_3 be different points in the same plane.

Let moment of the force "F" about point O_1 is MO,

$$MO_1 = F \times d_1$$

Let moment of the force "F" about point O_2 is MO,

$$MO_2 = F \times d_2$$

Let moment of the force "F" about point O_3 is MO,

MO= 0 x F

The given force produces a clockwise moment about point O_1 and anticlockwise moment about O_2.

A anticlockwise moment (↺) is treated as negative and an clockwise moment (↻) is treated as positive.

Note: The points O_1, O_2, O_3 about which the moments are calculated can also be called as moment centre.

Sign Convention for Moment of a Force

Clockwise moment positive and anticlockwise moment negative.

Let us determine moment of force 'F' about 'A' in the following cases.

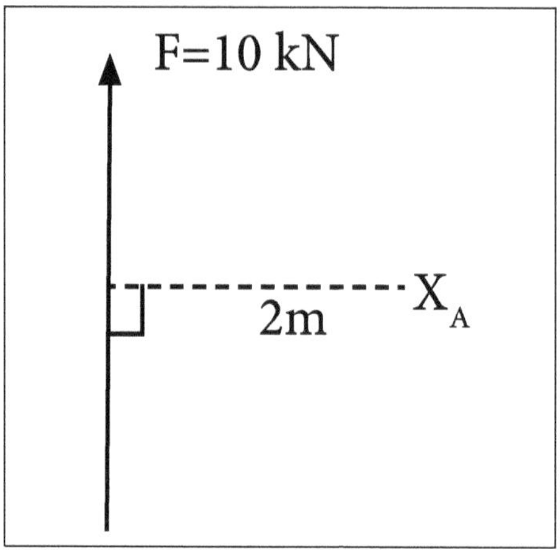

$M_A = + 10 \times 2$

$M_A = + 20$ kN m

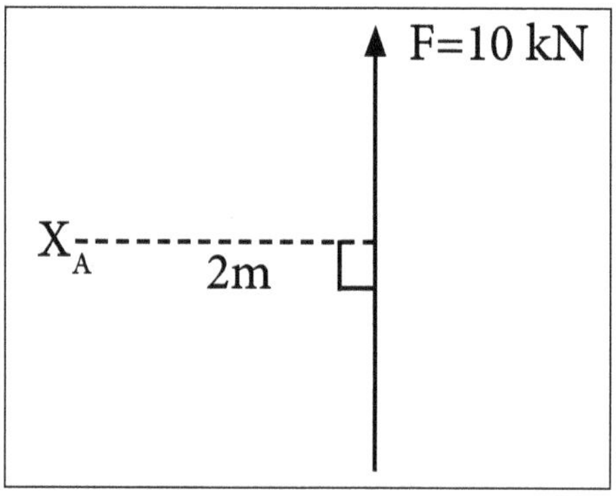

$M_A = - 10 \times 2$

$M_A = - 20$ kN m

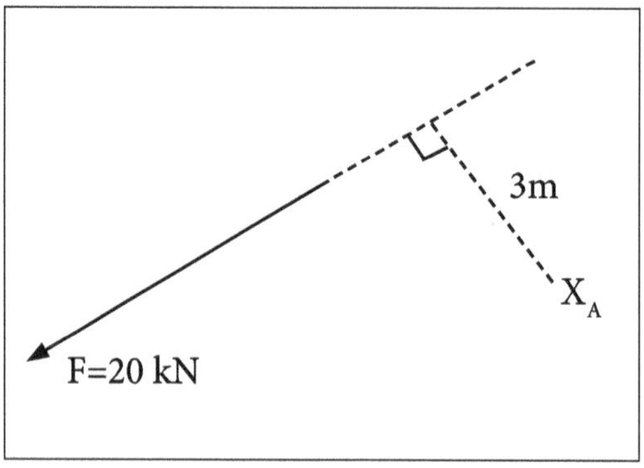

$M_A = - 20 \times 2$

$M_A = - 40$ kN m

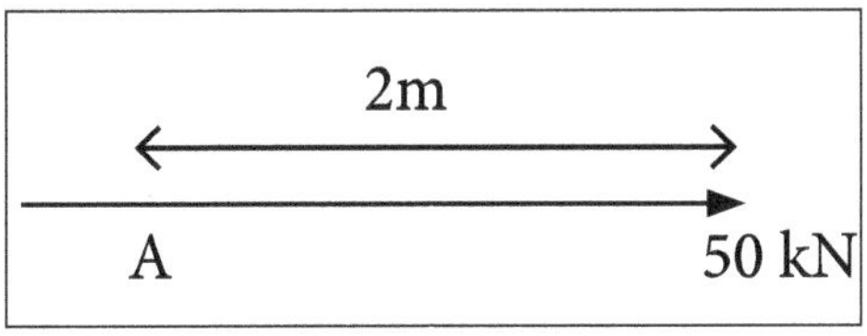

$M_A = 50 \times 0$

$M_A = 0$

Let us find moment of the force about A and B in the following.

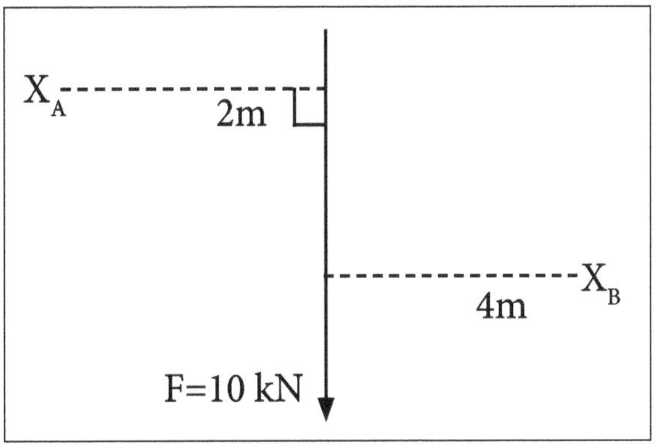

$M_A = +10 \times 2$

$M_A = +20$ kN m

$M_B = -10 \times 2$

$M_B = -20$ kN m

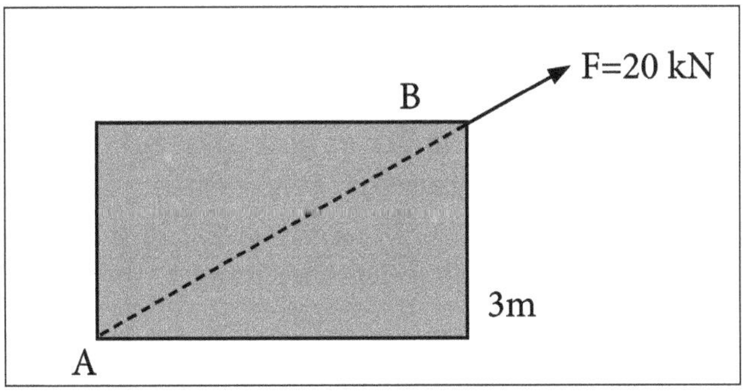

$M_A = 0$

$M_B = 0$

Calculation of Moment of a Force about a Point

Moment of a force about any point is given by the product of magnitude of force and perpendicular distance between the line of action of a force and the point about which moment is considered.

$$M_A = FL$$

Unit: Nm

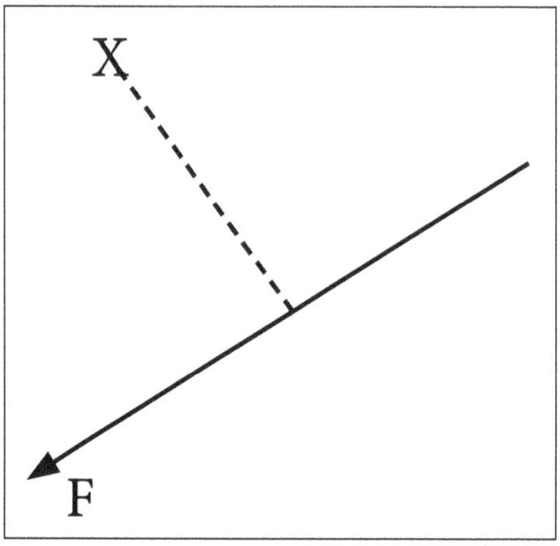

Applications

There are several real-world applications of these phenomena. One example is the use of air bags in automobiles. Air bags are used in automobiles because they are able to minimize the effect of the force on an object involved in a collision.

Air bags accomplish this by extending the time required to stop the momentum of the driver and passenger. When encountering a car collision, the driver and passenger tend to keep moving in accord with Newton's first law. Their motion carries them towards a windshield that results in a large force exerted over a short time in order to stop their momentum.

If instead of hitting the windshield, the driver and passenger hit an air bag, and then the time duration of the impact is increased. When hitting an object with some give such as an air bag, the time duration might be increased by a factor of 100. Increasing the time by a factor of 100 will result in a decrease in force by a factor of 100.

Fans of boxing frequently observe this same principle of minimizing the effect of a force by extending the time of collision. When a boxer recognizes that he will be hit in the head by his opponent, the boxer often relaxes his neck and allows his head to move backwards upon impact.

In the boxing world, this is known as riding the punch. A boxer rides the punch in order to extend the time of impact of the glove with their head. Extending the time results in decreasing the force and thus minimizing the effect of the force in the collision.

Merely increasing the collision time by a factor of ten would result in a tenfold decrease in the force.

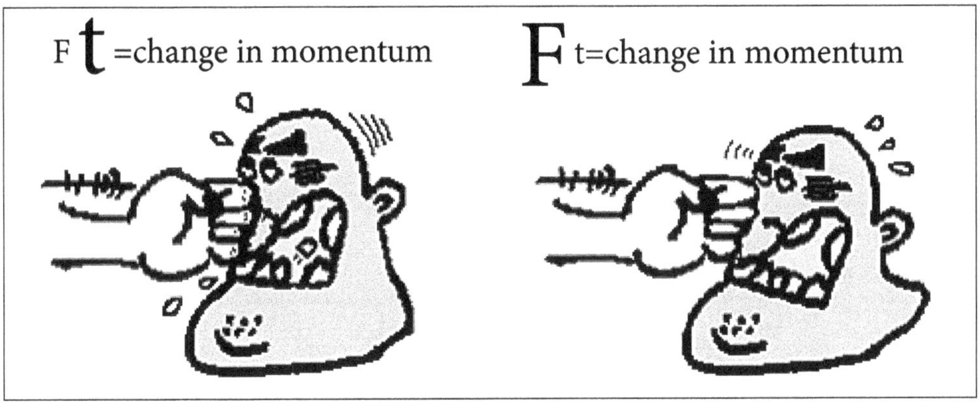

Riding the punch increases the time of collision and reduces the force of collision.

Nylon ropes are used in the sport of rock-climbing for the same reason. Rock climbers attach themselves to the steep cliffs by means of nylon ropes. If a rock climber should lose her grip on the rock, she will begin to fall.

In such a situation, her momentum will ultimately be halted by means of the rope, thus preventing a disastrous fall to the ground below.

The ropes are made of nylon or similar material because of its ability to stretch. If the rope is capable of stretching upon being pulled taut by the falling climber's mass, then it will apply a force upon the climber over a longer time period.

Extending the time over which the climber's momentum is broken results in reducing the force exerted on the falling climber. For certain, the rock climber can appreciate minimizing the effect of the force through the use of a longer time of impact.

Mountain climbers use nylon ropes to increase the stopping time and decrease the stopping force.

1.7.1 Equivalent Force Couple System

Every set of forces and moments has an equivalent force couple system. This is a pure moment and single force acting at a single point that is statically equivalent to the original set of moments and forces.

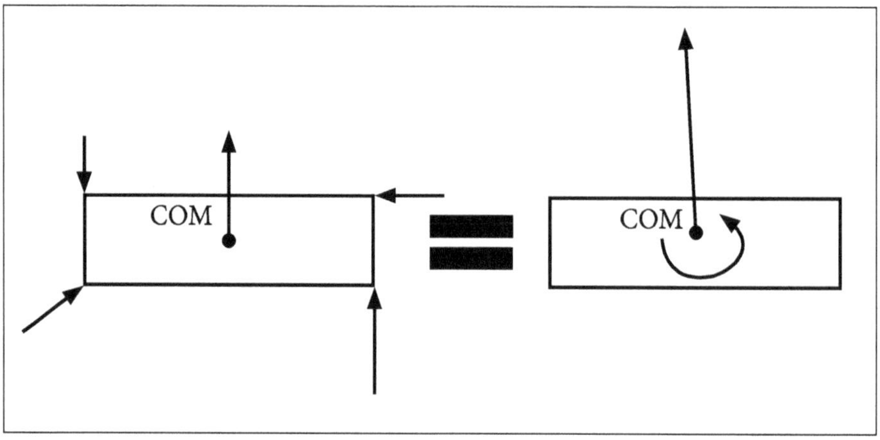

Any set of forces on a body can be replaced by a single force and a single couple acting, that is statically equal to the original set of forces and moments. This set of an equivalent force and a couple is called as the equivalent force couple system. To find the equivalent force couple system, we simply need to follow the steps below.

First, choose a point which is about to take the equivalent force couple system. Any point will work, but the point we choose will affect the final values we find for the equivalent force couple system. Traditionally this point will either be the some connection point for the body or center of mass of the body.

Next resolve all the forces not acting through that point to a force and a couple acting at the point we choose.

To find the "force" part of the equivalent force couple system, add all the force vectors together. This will give us the direction and the magnitude of the force in the equivalent force couple system.

To find the "couple" part of the equivalent force couple system, add together any moment vectors. This will give us the magnitude and direction of the pure moment in the equivalent force couple system.

1.7.2 Numerical Problems on Moment of Forces, Couples and on Equivalent Force couple System

1. Let us determine the resultant of the force system acting on the plate as shown in figure given below with respect to AB and AD.

Solution:

Given:

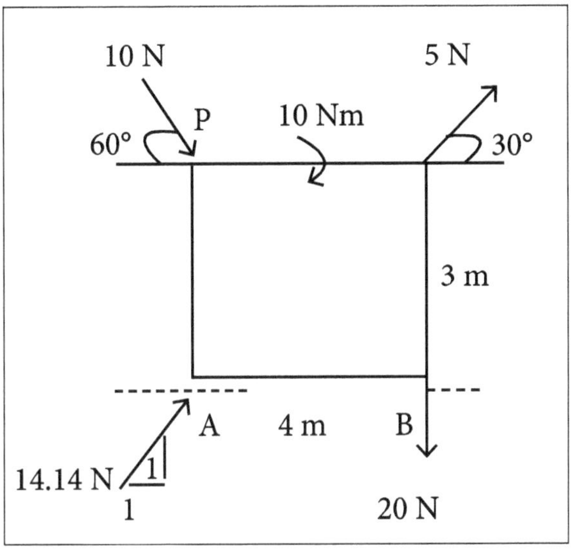

To find:

Resultant of force system acting on the plate,

$$\sum F_X = 5\cos\ 30^\circ + 10\ \cos\ 60^\circ + 14.14\ \cos\ 45^\circ$$

$$\sum F_X = 19.33\ \text{N}$$

$$\sum F\ = 5\sin\ 30^\circ - 10\ \sin\ 60^\circ + 14.14\ \sin\ 45^\circ$$

$$\sum F_y = -16.16\ \text{N}$$

$$R = \sqrt{\left(\sum F_x^2 + \sum F_y^2\right)}$$

$$R = 25.2\ \text{N}$$

$$\theta = \tan^{-1}\left(\frac{\sum F_y}{\sum F_x}\right)$$

$$\theta = \tan^{-1}\left(\frac{16.16}{19.33}\right)$$

$$\theta = 39.89^\circ$$

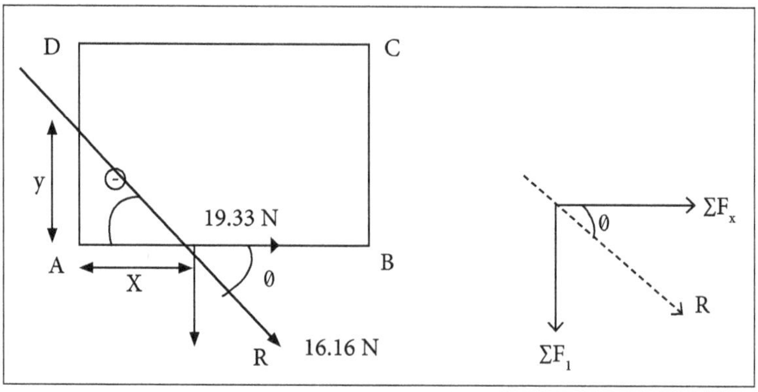

Tracing moments of forces about A and applying Varignon's principle of moments we get,

$$+16.16\ X = (20 \times 4) + (5\cos 30^\circ \times 3) - (5\sin 30^\circ \times 4) + 10 + (10\cos 60^\circ \times 3)$$

$$x = 107.99/16.16 = 6.683\ \text{m}$$

Also,

$$\tan 39.39^\circ = y / 6.83$$

$$y = 5.586\ \text{m.}$$

2. For the system of parallel forces shown below, let us determine the magnitude of the resultant and also its position from A.

Solution:

Given:

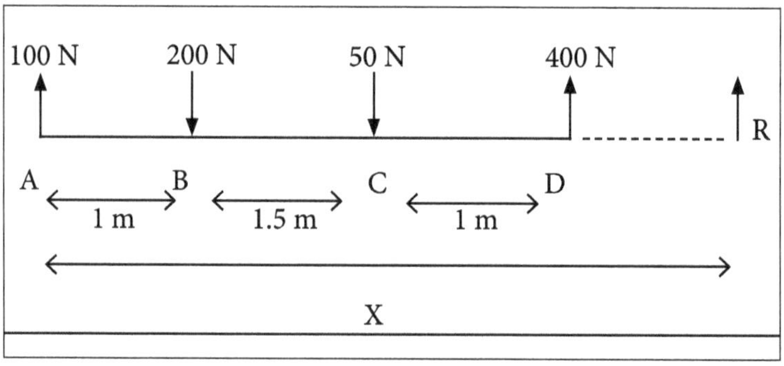

To find:

Magnitude and position from A,

$$\sum F_y = +100 - 200 - 50 + 400$$

$$\Sigma F_y = +250 \text{ N}$$

i.e.,

$$R = \sum F_y = 250 \text{ N} (\uparrow)$$

Since,

$$\sum F_x = 0$$

Taking moments of forces about A and applying Varignon's principle of moments we get,

$$-250 \text{ x} = (-400 \times 3.5) + (50 \times 2.5) + (200 \times 1) - (100 \times 0)$$

$$X = -1075 / -250 = 4.3\text{m}$$

3. The three like parallel forces 100 N and 300 N are acting as shown in figure below. If the resultant R=600 N is acting at a distance of 4.5 m from A, let us determine the magnitude of force F and position of F with respect to A.

Solution:

Given:

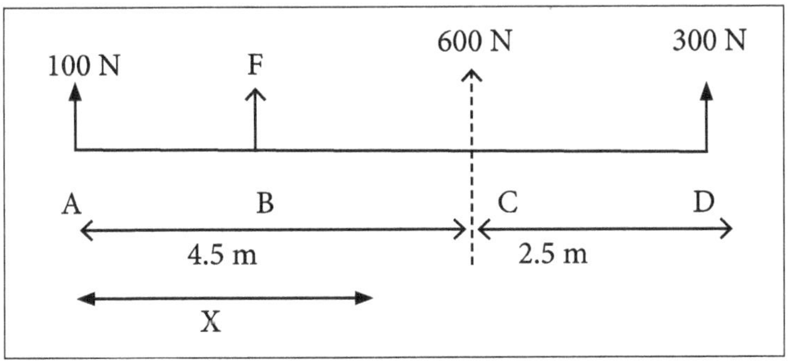

To find:

Magnitude and position of F

Let x be the distance from A to the point of application of force F

Here,

$$R = \sum F_y$$

600 = 100 + F + 300 F = 200 N

Taking moments of forces about A and applying Varignon's principle of moments we get,

$$(600 \times 4.5) = (300 \times 7) + F_x$$

$$200\,x = 600 \times 4.5 - 300 \times 7$$

$$X = 600/200 = 3\text{m from A}$$

4. A beam is subjected to forces as shown in the figure given below. Let us determine the magnitude, direction and the position of the resultant force.

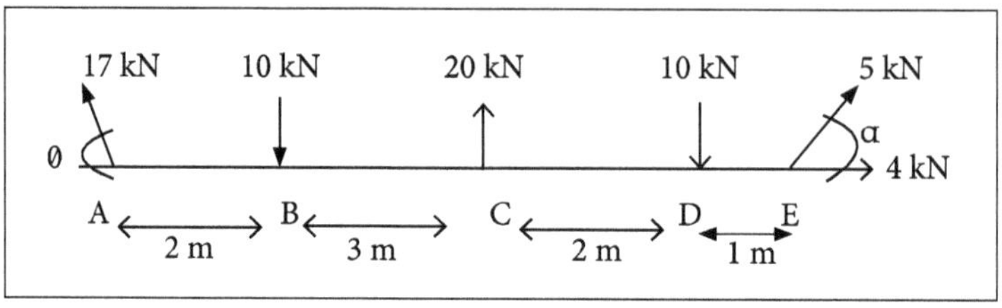

Solution:

Given:

$$\tan\theta = 15/8$$

$$\sin\theta = 15/17$$

$$\cos\theta = 8/17$$

$$\tan\alpha = 3/4$$

$$\sin\alpha = 3/5$$

$$\cos\alpha = 4/5$$

To find:

Magnitude, direction and position of the resultant force,

$$\sum F_x = 4 + 5\cos\alpha - 17\cos\theta$$

$$= 4 + 5 \times 4/5 - 17 \times 8/17$$

$$\sum F_x = 0$$

$$\sum F_y = 5\sin\alpha - 10 + 20 - 10 + 17\sin\theta$$

$$= 5 \times 3/5 - 10 + 20 - 10 + 17 \times 15/17$$

$$\sum F_y = 18 \text{ kN } (\uparrow)$$

Resultant force,

$$R = (\sum F_x)^2 + (\sum F_y)^2 = 0 + 182$$

$$R = 18 \text{ kN} (\uparrow)$$

Let x = Distance from A to the point of application R

Taking moments of forces about A and applying Varignon's theorem of moments we get,

$$-18 \text{ x} = -5 \text{ x } \sin \alpha \text{ x} 8 + 10 \text{ x } 7 - 20 \text{ x } 5 + 10 \text{ x } 2$$

$$= -3 \text{ x } 8 + 10 \text{ x } 7 - 20 \text{ x } 5 + 10 \text{ x } 2$$

$$x = -34 / -18 = 1.89 \text{ m from A (towards left).}$$

2

Analysis of Concurrent Force Systems

2.1 Resultants, Equilibrium and Composition of Forces

Resultant of a force system is a force or a couple that will have the same effect to the body, both in translation and rotation, if all the forces are removed and replaced by the resultant.

The equation involving the resultant of force system are the following:

- $R_x = \sum F_x = F_{x1} + F_{x2} + F_{x3} + ...$ The x-component of the resultant is equal to the summation of forces in the x-direction.
- $R_y = \sum F_y = F_{x1} + F_{x2} + F_{x3} + ...$ The y-component of the resultant is equal to the summation of forces in the y-direction.
- $R_z = \sum F_z = F_{x1} + F_{x2} + F_{x3} + ...$ The z-component of the resultant is equal to the summation of forces in the z-direction.

Resultant of Coplanar Concurrent Force System

The line of action of each forces in coplanar concurrent force system are on the same plane. All of these forces meet at a common point, thus concurrent. In x-y plane, the resultant can be found by the following formulas.

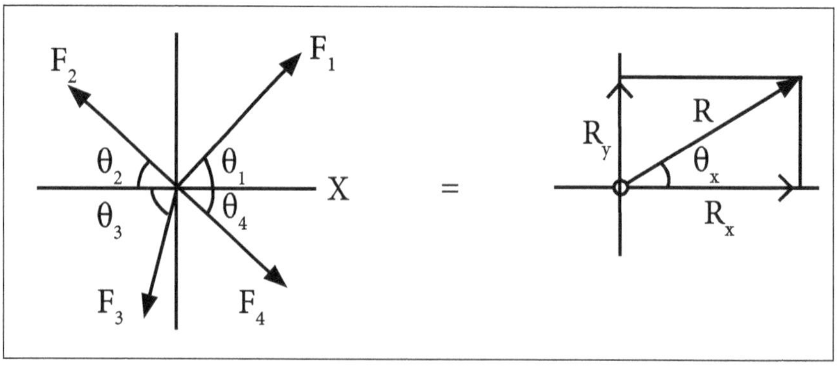

$$R_x = \sum F_x$$

$$R_y = \sum F_y$$

$$R = \sqrt{\left(R_{x2} + R_{y2} \right)}$$

$$\tan \theta_x = R_y / R_x$$

Resultant of Spatial Concurrent Force System

Spatial concurrent forces (forces in 3-dimensional space) meet at a common point but do not lie in a single plane. The resultant can be found as follows:

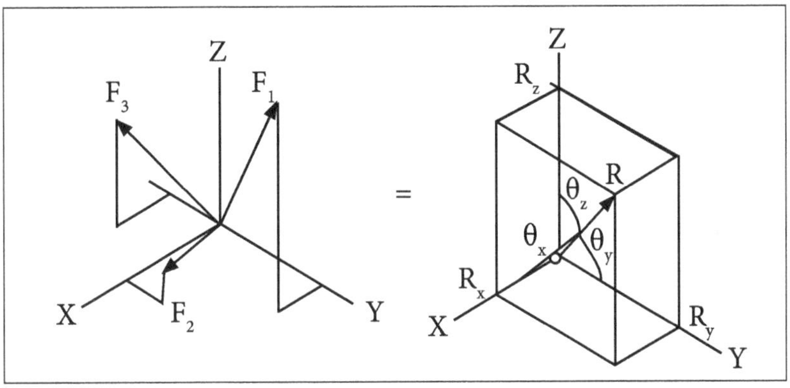

$$R_x = \sum F_x$$

$$R_y = \sum F_y$$

$$R_z = \sum F_z$$

$$R = \sqrt{\left(R_x^2 + R_y^2 + R_z^2 \right)}$$

Direction Cosines

$$\cos\theta_x = R_x / R$$

$$\cos\theta_y = R_y / R$$

$$\cos\theta_z = R_z / R$$

Vector Notation of the Resultant

$$R = \sum F$$

$$R = \left(\sum F_x\right)i + \left(\sum F_y\right)j + \left(\sum F_z\right)k$$

$$R = R_x i + R_y j + R_z k$$

Where,

$$R_x = \sum F_x$$

$$R_y = \sum F_y$$

$$R_z = \sum F_z$$

$$R = \sqrt{(R_x^2 + R_y^2 + R_z^2)}$$

Equilibrium

The body is said to be in equilibrium if the resultant of all forces acting on it is zero. There are two major types of static equilibrium, namely, translational equilibrium and rotational equilibrium.

Formulas Concurrent force system,

$$\sum F_x = 0$$

$$\sum F_y = 0$$

Parallel Force System,

$$\sum F = 0$$

$$\sum M_O = 0$$

Non-Concurrent Non-Parallel Force System,

$$\sum F_x = 0$$

$$\sum F_y = 0$$

$$\sum M_O = 0$$

Force system	Equations of equilibrium	Remarks
Spatial general	$\sum F_{ix} = 0$	$\sum F_{ix}$ is algebraic sum of x components
	$\sum F_{iy} = 0$	Of the forces; similarly for other axes

	$\sum F_{iz} = 0$	
	$\sum M_{ix} = 0$	$\sum M_{ix}$ is algebraic sum of the moments
	$\sum M_{iy} = 0$	Forces of the system acts about x-axis;
	$\sum M_{iz} = 0$	Similarly for other axes
Spatial parallel	$\sum F_{iz} = 0$	Forces line of action are parallel to z-axis
	$\sum F_{ix}$	
	$\sum M_{iy} = 0$	
Spatial concurrent	$\sum F_{ix} = 0$	
	$\sum F_{iy} = 0$	
	$\sum F_{iz} = 0$	
Coplanar	$\sum F_{ix} = 0$	Forces lie in xy plane
	$\sum F_{iy} = 0$	
	$\sum M_{iz} = 0$	
Coplanar parallel	$\sum F_{iy} = 0$	Forces lie in xy plane
	$\sum M_{iz} = 0$	Lines of action are parallel toy-axis
Coplanar concurrent	$\sum F_{ix} = 0$	Forces lie in xy plane
	$\sum F_{iy} = 0$	
Collinear	$\sum F_{ix} = 0$	Line of action is x-axis

Composition of Forces

The reduction of a given system of forces to the simplest system that will be its equivalent is known as the problem of composition of forces.

Composition of Forces by Resolution

- The components of each force in the system in two mutually perpendicular directions are found.
- The components in each direction are algebraically added together to obtain the two components.
- These two component forces which are mutually perpendicular are combined to obtain the resultant force.

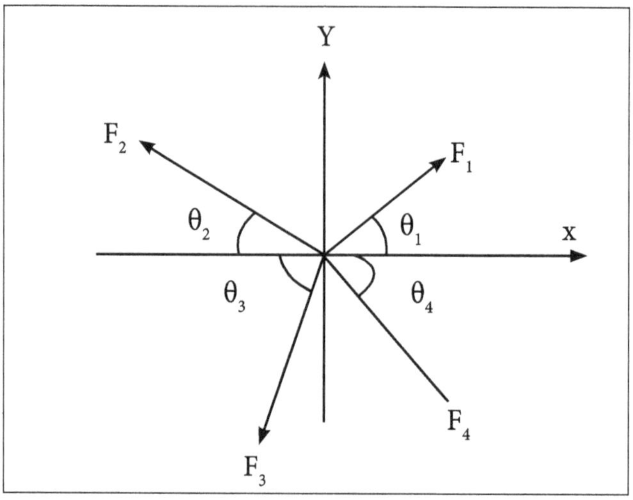

Algebraic sum of the components of forces in X direction.

$$\sum F_x = F_1 \cos\theta_1 - F_2 \cos\theta_2 - F_3 \cos\theta_3 + F_4 \cos\theta_4$$

Algebraic sum of the components of forces in Y direction,

$$\sum F_y = F_1 \sin\theta_1 + F_2 \sin\theta_2 - F_3 \sin\theta_3 - F_4 \sin\theta_4$$

Now the system of forces is equivalent to two mutually perpendicular forces,

$$\sum F_X \text{ and } \sum F_Y$$

$$R = \sqrt{\sum F_X^2 + \sum F_Y^2}$$

$$\theta = \tan^{-1}\left(\frac{\sum F_Y}{\sum F_X}\right)$$

2.1.1 Definition of Resultant

Resultant Force

It is possible to find a single force which will have the same effect as that of a number of forces acting on a body. Such a single force is known as resultant force.

The process of finding out the resultant force is called as composition of forces.

The resultant of a system of forces is a simpler system of forces which has the same component of force in any direction and the same moment about any axis or point as the given system.

The dynamic effect of the resultant force acting on a rigid body will be the same as the effect of the given system of forces. If all forces are concurrent, the simplest form of the resultant is a single force.

If all forces are parallel or coplanar they may be reduced to a single force or a single couple.

Resultant of a General Coplanar Force System

The resultant R of a general coplanar system of forces may be:

- A single force.
- A couple in the plane of the system or in a parallel plane or Zero.

The resultant R corresponds to the vector sum of the forces of the system. Its x and y components are,

$$R_x = \sum F_x$$
$$R_y = \sum F_y$$

Where, $\sum F_x$ and $\sum F_y$ are the algebraic sums of the x and y components, respectively, of the forces of the system.

The magnitude of R is given by,

$$R = \sqrt{\sum F_x^2 + \sum F_y^2}$$

And the angle θ that it makes with the x axis is given by,

$$\theta = \tan^{-1}\left(\frac{\sum F_Y}{\sum F_X}\right)$$

The line of action of R is found from,

$$R_d = \sum M_O$$

Where,

O = any moment center in the plane.

d = Perpendicular distance from the moment center O to the resultant R.

$\sum M_O$ = the algebraic sum of the moments of the forces of the system with respect to O.

R_d = the moment of R with respect to O.

It should be noted that even if R = 0, a couple may exist with a magnitude equal to $\sum M_O$.

Resultant of a General Spatial Force System

The resultant of a general spatial force system is a force R and a couple C.

Where,

$R = \sum F$, the vector sum of all the forces of the system.

$C = \sum M$, the vector sum of the moments of all the forces of the system.

The value of R is independent of the co-ordinate system but the value of C depends on the center of moments chosen.

Erect an x-y-z co-ordinate system with origin O placed at some selected point inside or near the body. The origin O of the system will serve as the center of moments.

The x, y and z components of R are,

$$R_x = \sum F_x,\ R_y = \sum F_y,\ R_z = \sum F_z$$

Where, $\sum F_x$, $\sum F_y$ and $\sum F_z$ are the algebraic sums of the x, y and z components, respectively, of the forces of the system.

The x, y and z components of C are,

$$C_x = \sum M_x,\ C_y = \sum M_y,\ C_z = \sum M_z$$

Where, $\sum M_x$, $\sum M_y$ and $\sum M_z$ are the algebraic sums of the moments of the forces of the system about the x, y and z axes respectively.

The magnitudes of R and C are given by,

$$R = \sqrt{\left(\sum F_x\right)^2 + \left(\sum F_y\right)^2 + \left(\sum F_z\right)^2}$$

And,

$$C = \sqrt{\left(\sum M_x\right)^2 + \left(\sum M_y\right)^2 + \left(\sum M_z\right)^2}$$

If θ_x, θ_y and θ_z are the angles that R makes with the x, y and z axes respectively and α_x, α_y and α_z are the angles that C makes with the x, y and z axes respectively, then,

$$\cos\theta_x = \left(\sum F_x\right)/R \quad \cos\theta_y = \left(\sum F_y\right)/R \quad \cos\theta_z = \left(\sum F_z\right)/R$$

$$\cos\alpha_x = \left(\sum M_x\right)/C \quad \cos\alpha y = \left(\sum M_y\right)/C \quad \cos\alpha z = \left(\sum M_z\right)/C$$

Derivation of R = ΣF and C = ΣM. The derivation of the formulas employs the idea of resolving a force into a force and a couple.

Problems

1. Five forces act on a bolt 'B' is shown in figure. Let us determine the resultant of the forces on.

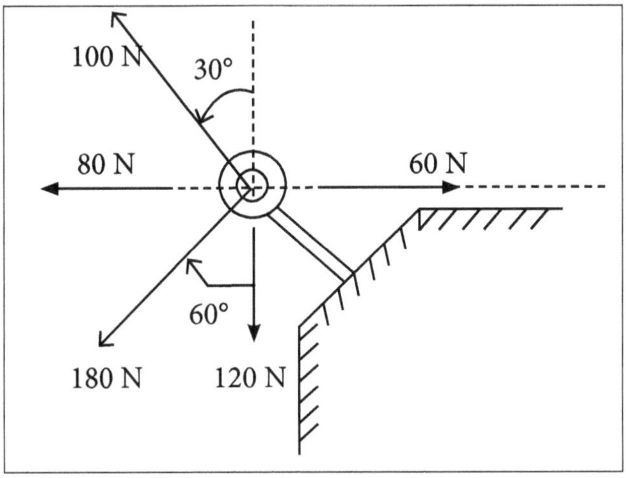

Solution:

Given:

S.No.	Force (kN)		Horizontal (H) Component	Vertical(V) Component
1.	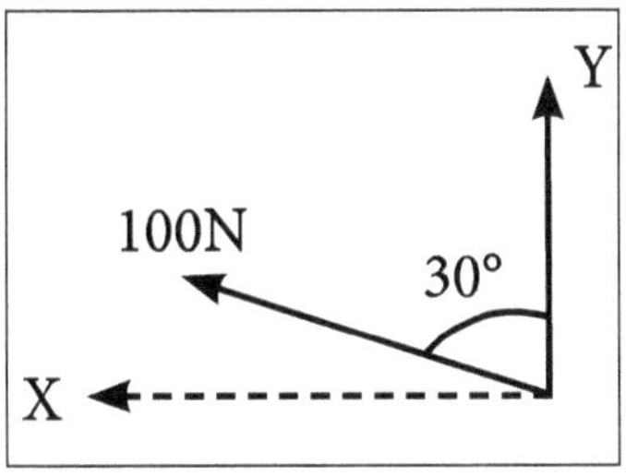		−100 cos60° = − 50	100 sin60° = 86.6

2.	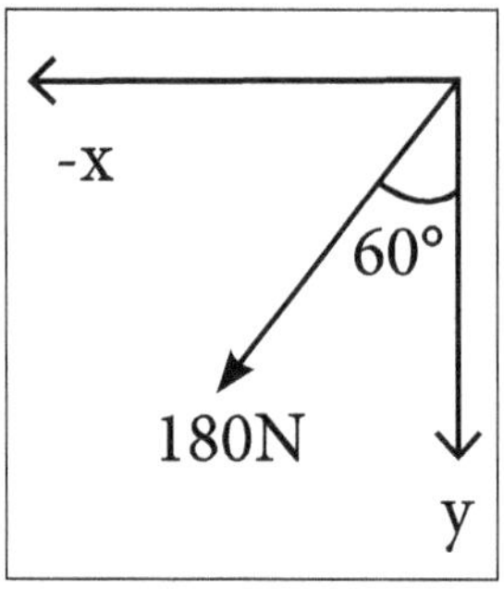	$\overleftarrow{80\text{N}}$	−80N	0
3.			−180cos60° = −90	180 sin 60° =155.8
4.		↓ 120N	−120	0
5.		→ 60N	60	0

Formula to be used:

$$R = \sqrt{\left(\sum H\right)^2 + \left(\sum V\right)^2}$$

$$\tan\alpha = \frac{\sum V}{\sum H}$$

Total $\sum H = -160$ $\sum V = 62.4$

$$R = \sqrt{\left(\sum H\right)^2 + \left(\sum V\right)^2}$$

$$R = \sqrt{25600 + 389376}$$

$$R = 171.7 \text{ N}$$

$$\tan\alpha = \frac{\sum V}{\sum H}$$

$$\tan\alpha = \frac{62.4}{160}$$

$$\alpha = 21.3^\circ$$

Result:

Resultant force R = 171.7 N

Direction of resultant $\alpha = 21.3°$.

2. A system of concurrent forces 50 kN, 100 kN and 125 kN act at a point O (0, 0, 0) and the forces are directed respectively through the points whose co-ordinates are P(3, 2, 6), Q(7, 8, - 4) and R(- 4, 4, - 6). Let us determine the resultant and its direction.

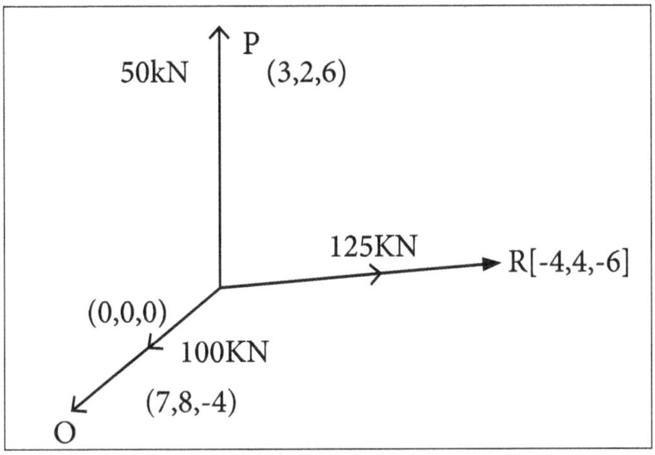

Solution:

Given:

Forces = 50 kN, 100 kN and 125 kN

Co-ordinates = P(3, 2, 6), Q(7, 8, - 4) and R(- 4, 4, - 6)

Formula to be used:

$$\vec{F} = F x \hat{n}PQ$$

$$\hat{n}PQ = \frac{(XQ-XP)i+(YQ-YP)j+(ZQ-ZP)k}{\sqrt{(XQ-XP)^2+(YQ-YP)^2+(ZQ-ZP)^2}} \quad \hat{n}PQ = \frac{PQ}{|PQ|}$$

$$\hat{n}PQ = \frac{(7-3)i+(8-2)j+(-4-6)k}{\sqrt{(7-3)^2+(8-2)^2+(-4-6)^2}}$$

$$\hat{n}PQ = \frac{4i+6j+10k}{\sqrt{16^2+36^2+100^2}}$$

$$\hat{n}PQ = \frac{4i+6j+10k}{107.4}$$

F = 50 + 100 + 125 = 275 KN

$$F_X = 8.25 \text{ kN}$$
$$F_Y = 13.75 \text{ kN}$$
$$F_2 = -\ 25.5 \text{ kN.}$$

Direction:

$$\theta_x = \cos^{-1}\left(\frac{F_X}{F}\right) = 88.2^\circ$$

$$\theta_y = \cos^{-1}\left(\frac{F_y}{F}\right) = 89.9^\circ$$

$$\theta_z = \cos^{-1}\left(\frac{F_z}{F}\right) = -84.8^\circ$$

3. Let us determine the resultant of the concurrent force system shown in figure:

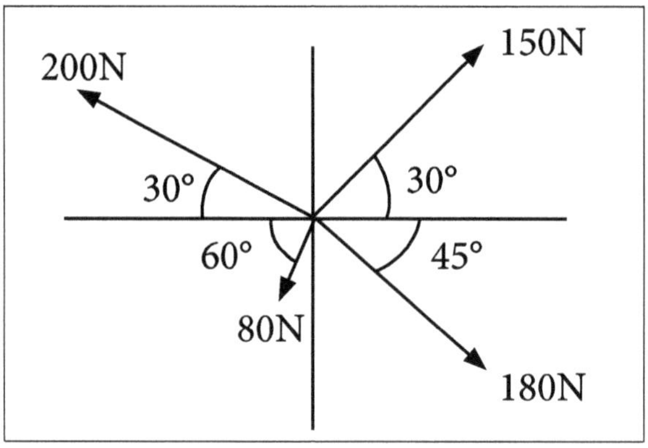

Solution:

Given:

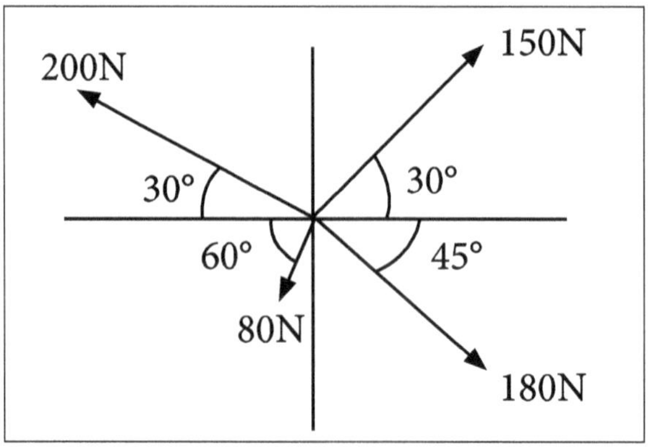

Formula to be used:

$$R = \sqrt{F_X^2 + F_Y^2}$$

$$Q = \tan^{-1}\left(\frac{F_Y}{F_X}\right)$$

Resolve all the forces into horizontal and vertical component's as shown in the figure:

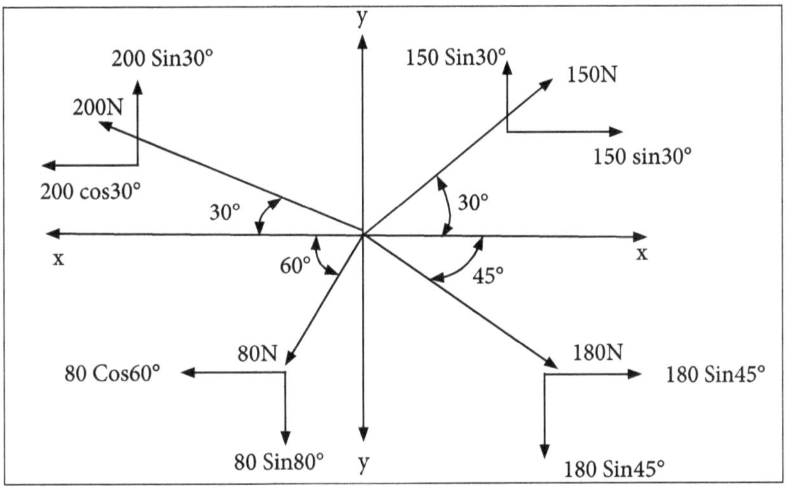

$$\sum F_x = 0$$

$$= +\ 150 \cos 30° + 180 \cos 45° - 200 \cos 30^0 - 80 \cos 60°$$

$$\sum F_X = 129.9 + 127.27 - 173.20 - 40$$

$$= 257.17 - 213 \sum F_X = 43.97 \text{ N.}$$

$$\sum F_Y = 0$$

$$= +\ 150 \sin 30 + 200 \sin 30 - 80 \sin 60 - 180 \sin 45$$

$$= 75 + 100 - 69.28 - 127$$

$$= 175 - 196.28$$

$$F_Y = -21.28 \text{ N.}$$

Resultant forces acting in the origin is given by,

$$R = \sqrt{F_X^2 + F_Y^2}$$

$$R = \sqrt{(43)^2 + (-21)^2}$$

$$R = 47\ \text{N}$$

$$Q = \tan^{-1}\left(\frac{F_Y}{F_X}\right)$$

$$Q = \tan^{-1}\left(\frac{-21}{43}\right) = -26°$$

Result:

The resultant force = 47 N

$\theta = -26°$.

4. The truck shown is to be towed using two ropes. Let us determine the magnitudes of forces F_A and F_B acting on each rope in order to develop a resultant force of 950 N directed along the positive X-axis.

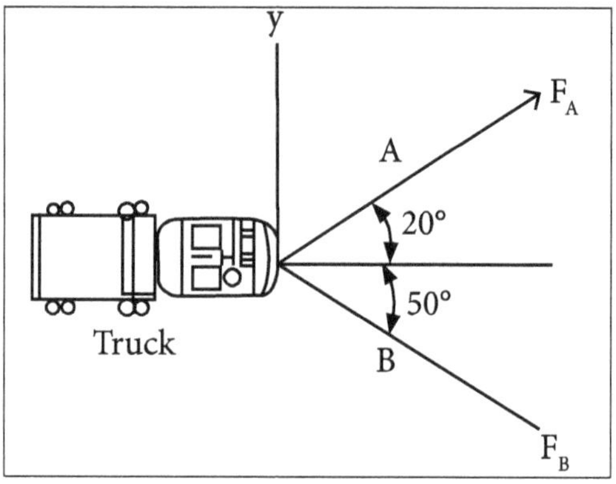

Solution:

Given:

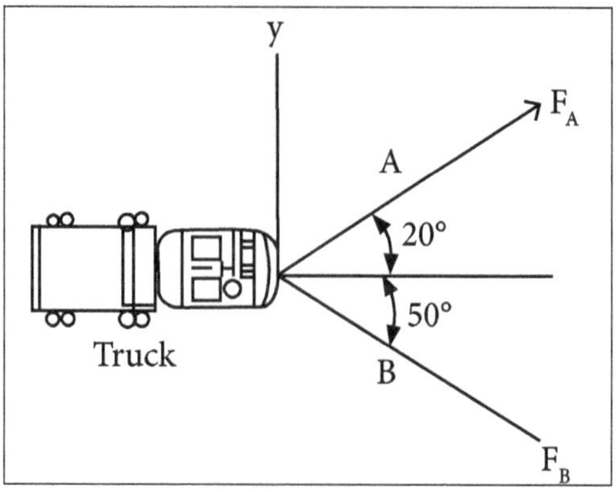

Resolving the forces.

$$\sum H = F_A \cos 20° + F_B \cos 50°$$

$$= 0.939\ F_A + 0.642\ F_B \quad ...(1)$$

$$\sum v = F_A \sin 20° + F_B \text{ Sin } 50°$$

$$= 0.342\ F_A + 0.766\ F_B \quad ...(2)$$

Given,

R = 950 *N* is acting along the positive direction of x-axis.

$$\Rightarrow \sum H = 950\ \sum V = 0.$$

$$950 = 0.939\ F_A + 0.642\ F_B \quad ...(3)$$

$$\Rightarrow 0 = 0.342\ F_A + 0.766\ F_B$$

$$-\ 0.342\ F_A = +\ 0.\ 766\ F_B$$

$$F_A = -\frac{0.766}{0.342} F_B$$

$$F_4 = -2.239\ F_B \quad ...(4)$$

Put (4) in (3),

$$950 = 0.939\ (-\ 2.239\ F_B) + 0.624\ F_B$$

$$950 = 1.478\ F_B$$

$$F_B = 642.76\ \text{N}$$

$$F_A = -\ 2.239\ F_B$$

$$F_A = -\ 2.239(642.76)$$

$$F_A = -\ 1439.13\ \text{N}$$

2.1.2 Composition of Coplanar and Concurrent Force System

Coplanar forces means the forces in a plane. When many forces act on a body, then they are called as force system or a system of forces. A system in which all the forces lie in the same plane, is known as coplanar force system.

If the forces are having common line of action, then they are known as collinear where-as if the forces intersect at a common point, then they are known as concurrent.

A force system may be coplanar or non-coplanar. If in a system all the forces lie on the same plane then the force system is known as coplanar.

But if in a system all the forces lie in different planes, then the force system is known as non- coplanar. Hence a force system is classified as shown in figure:

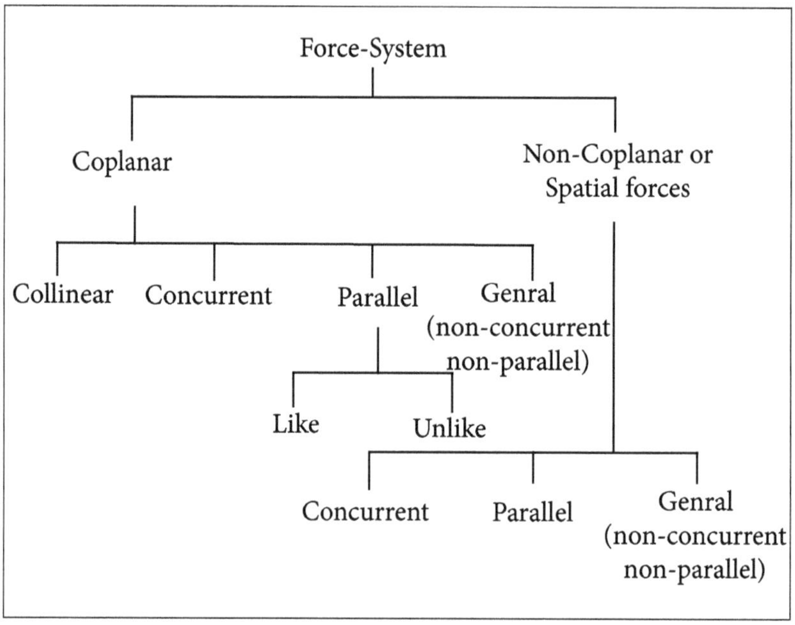

Classification of force system.

The coplanar forces may be:

- Collinear.
- Concurrent.
- Parallel.
- Non-concurrent, non-parallel.

Collinear

A three forces F_1, F_2 and F_3 act in a plane. These 3 forces are in the same line i.e., these three forces are having a common line of action.

This system of forces is known as coplanar collinear force system. Hence, in coplanar collinear system of forces, all the forces act in the same plane and have a common line of action as shown in figure:

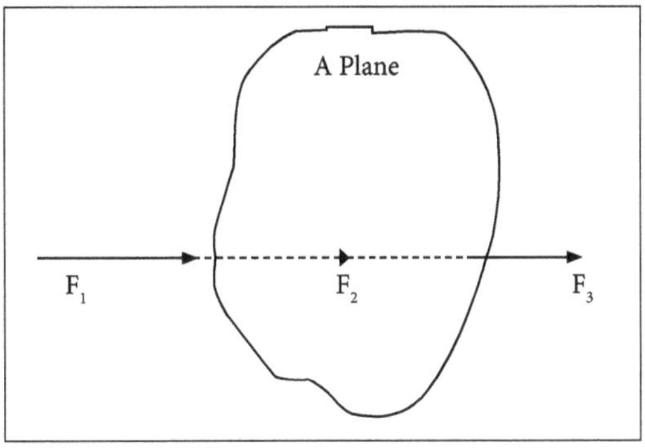

Coplanar collinear system.

Concurrent

Let three forces F_1, F_2 and F_3 act in a plane and intersect or meet at a common point O. This system of forces is known as coplanar concurrent force system. Hence, in coplanar concurrent system of forces, all the forces act in the same plane and they intersect at a common point as shown in figure.

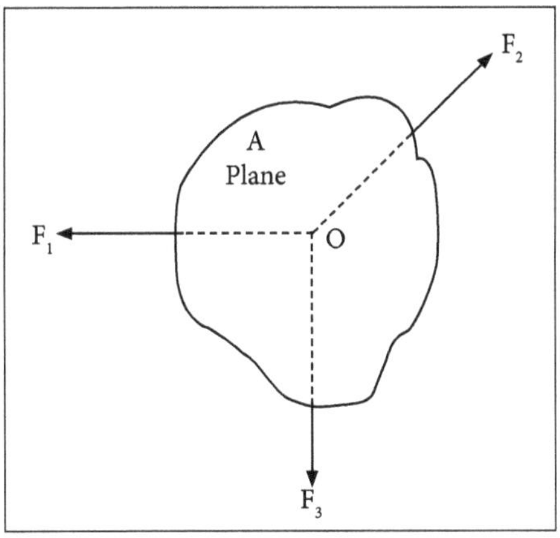

Coplanar concurrent force system.

Parallel

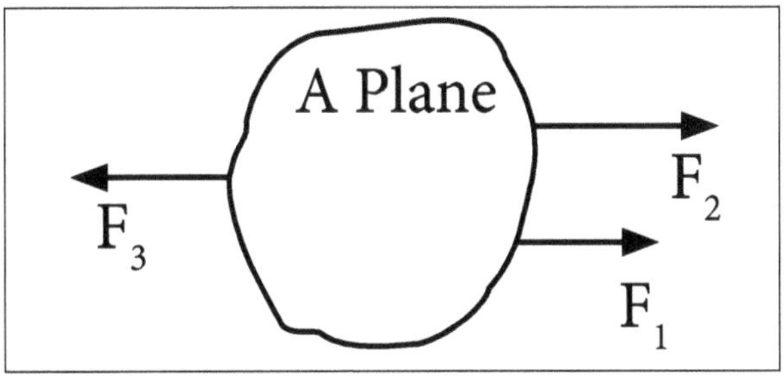

Parallel force system.

Figure shows three forces F_1, F_2 and F_3 acting in a plane and these forces are parallel. This system of forces is known as coplanar parallel force system. Hence in coplanar parallel system of forces, all the forces act in the same plane and are parallel.

Coplanar Non-concurrent Non-parallel

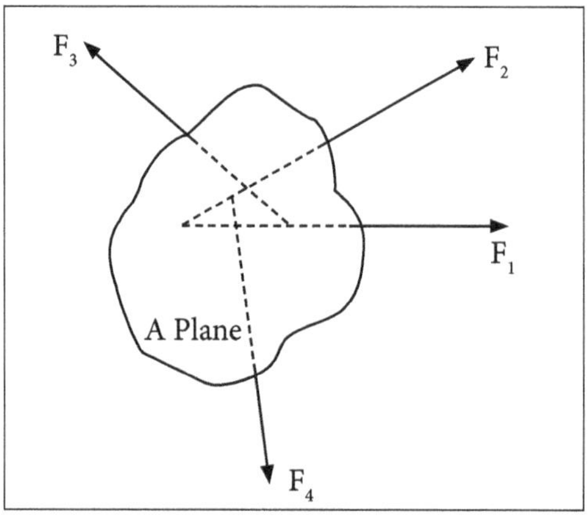

Coplanar Non-concurrent Non-parallel system.

Figure shows four forces F_1, F_2, F_3 and F_4 acting in a plane. The lines of action of these forces lie on the same plane but they are neither parallel nor meet or intersect at a common point.

This system of forces is known as coplanar non-concurrent non-parallel force system. Hence in coplanar non-concurrent non-parallel system of forces, all the forces act in the same plane but the forces are neither parallel nor meet at common point. This force system is also known as general system of forces.

Coplanar Concurrent Force System

A concurrent force system contains forces whose line of action meet at one point. Forces may be tensile (pulling) or forces may be compressive (pushing).

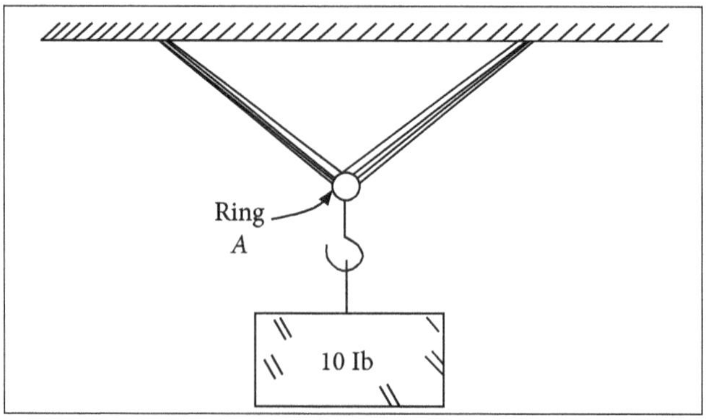

Two wires supporting a load.

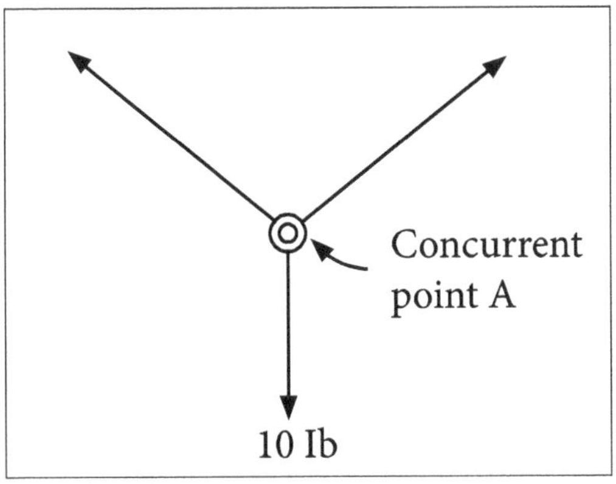

Forces acting on ring.

2.1.3 Parallelogram Law of Forces and Principle of Resolved Parts

Resolution of a Force

The process of substituting a force by its components so that the net effect on the body remains the same is known as resolution of a force.

For each force, there exists an infinite number of possible sets of components.

Suppose a force is to be resolved into two components. Then:

- When one of the components is known, the second component can be obtained by applying the triangle rule.
- When the line of action of each component is known, the magnitude and the sense of the components are obtained by parallelogram law.

Principle of Resolution

The algebraic sum of the resolved parts of a number of forces in the given direction is equal to the resolved part of their resultant in the same direction.

Composition of Forces

The process of finding out the resultant force of a number of given forces is called the composition/compounding of forces.

Parallelogram Law of forces

The parallelogram law of forces is used to calculate the resultant of two forces acting at a point in a plane.

It states, "If two forces, acting at a point be represented in magnitude and direction by the two adjacent sides of a parallelogram, then their resultant is expressed in terms of magnitude and direction by the diagonal of the parallelogram passing through that point."

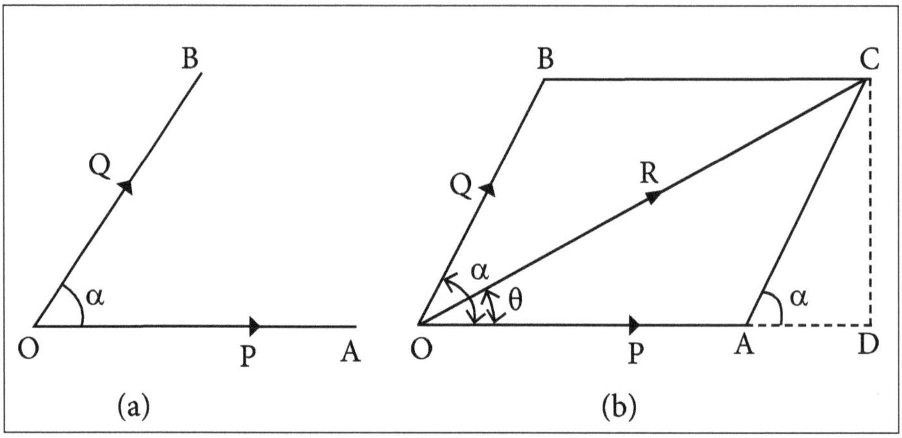

Parallelogram law of forces.

Let two forces P and Q act at a point O as shown in figure (a). The force P is represented in magnitude and direction by OA whereas, the force Q is represented in magnitude and direction by OB.

Let the angle between the two forces be 'a'. The resultant of these two forces can be obtained in direction and magnitude by the diagonal of the parallelogram of which OA and OB are two adjacent sides. Hence draw the parallelogram with two adjacent sides OA and OB as shown in figure (b).

The resultant R is represented by OC in direction and magnitude.

Magnitude of Resultant (R)

From point C draw CD perpendicular to OA produced.

Let,

$$\alpha = \text{Angle between two forces P and Q} = \text{AOB}$$

Now,

$$\text{DAC} = \llcorner\text{AOB} \quad \text{(Corresponding angles)}$$

In parallelogram OACB, AC is parallel and equal to OB.

$$AC = Q.$$

In triangle ACD,

$$AD = AC\ \cos\ \alpha = Q\ \cos\ \alpha$$

And,

$$CD = AC \ \sin \alpha = Q \ \sin\alpha.$$

In triangle OCD,

$$OC^2 = OD^2 + DC^2$$

But,

$$OC = R, \ OD = OA + AD = P + Q \ \cos \alpha$$

And,

$$DC = Q \ \sin \alpha$$

$$\therefore R^2 = (P + Q \ \cos \alpha)^2 + (Q \ \sin \alpha)^2$$

$$R^2 = P^2 + Q^2 \cos^2 \alpha + \ 2PQ \ \cos \alpha + \ Q^2 \ \sin^2 \alpha$$

$$R^2 = p^2 + Q^2 \left(\cos^2 \alpha + \sin^2 \ \alpha\right) + 2PQ \ \cos \alpha$$

$$R^2 = P^2 + Q^2 + 2PQ \ \cos \ \alpha\left(\because \cos^2 \ \alpha + \sin^2 \ \alpha = 1\right)$$

$$\therefore R\sqrt{p^2 + Q^2 + 2PQ\cos\alpha}$$

The above equation gives the magnitude of resultant force R.

Direction of Resultant

Let,

θ = The angle made by resultant with OA.

Then from triangle OCD,

$$\tan\theta = \frac{CD}{OD} = \frac{Q\sin\alpha}{P + Q\sin\alpha}$$

$$\theta = \tan^{-1}\left(\frac{Q\sin\alpha}{P + Q\sin\alpha}\right)$$

The above equation gives the direction of resultant (R).

The direction of resultant (R) can also be obtained by using sine rule [In triangle OAC, OA = P, AC = Q, OC = R, angle OAC $=(180°-\alpha)$ angel ACO $=180°-[\theta+180°-\alpha]=(\alpha-\theta)$].

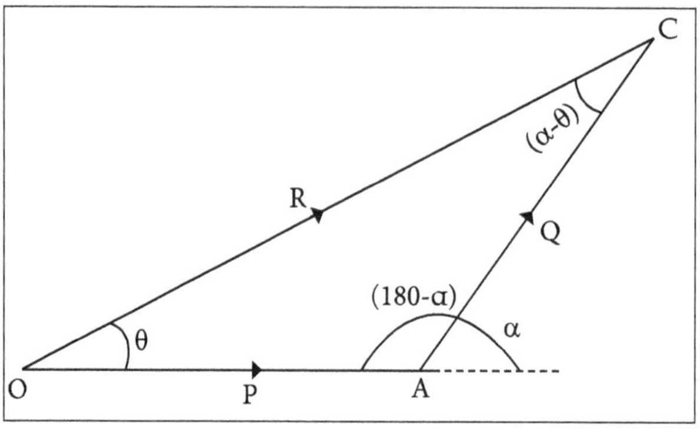

$$\frac{\sin\theta}{AC}=\frac{\sin(180-\alpha)}{OC}=\frac{\sin(\alpha-\theta)}{OA}$$

$$\frac{\sin\theta}{Q}=\frac{\sin(180-\alpha)}{R}=\frac{\sin(\alpha-\theta)}{P}$$

Two cases are important,

1st case: If the two forces P and Q act at right angles, then $\alpha = 90°$, we get the magnitude of resultant and the direction of resultant is obtained as,

$$R=\sqrt{P^2+Q^2+2PQ\cos\alpha}=\sqrt{P^2+Q^2+2PQ\cos 90°}$$

$$=\sqrt{P^2+Q^2} \qquad (\because \cos 90°=0)$$

2nd case: The two forces P and Q are equal and are acting at an angle α between them. Then the magnitude and direction of resultant is given as,

$$R=\sqrt{P^2+Q^2+2PQ\cos\alpha}=\sqrt{P^2+P^2+2P\times P\times\cos\alpha} \qquad (\because P=Q)$$

$$=\sqrt{2P^2+2P^2\cos\alpha}$$

$$=\sqrt{2P^2(1+\cos\alpha)}$$

$$=\sqrt{2P^2\times 2\cos^2\frac{\alpha}{2}} \qquad \left(\because 1+\cos\alpha=2\cos^2\frac{\alpha}{2}\right)$$

$$= \sqrt{4P^2 \cos^2 \frac{\alpha}{2}} = 2P \cos \frac{\alpha}{2}$$

$$\theta = \tan^{-1}\left(\frac{Q \sin \alpha}{P + Q \cos \alpha}\right) = \tan^{-1} \frac{P \sin \alpha}{P + P \cos \alpha} \quad (\because P = Q)$$

$$\tan^{-1} \frac{P \sin \alpha}{P(1 + \cos \alpha)} = \tan^{-1} \frac{\sin \alpha}{1 + \cos \alpha} = \tan^{-1} \frac{\sin \frac{\alpha}{2}}{\cos \frac{\alpha}{2}}$$

$$= \tan^{-1}\left(\tan \frac{\alpha}{2}\right) \frac{\alpha}{2} = \tan^{-1} \frac{2 \sin \frac{\alpha}{2} \cos \frac{\alpha}{2}}{2 \cos^2 \frac{\alpha}{2}} \quad \left(\because \sin \alpha = 2 \sin \frac{\alpha}{2} \cos \frac{\alpha}{2}\right)$$

2.1.4 Numerical Problems on Composition of Coplanar Concurrent Force Systems

1. A plate ABCD in the shape of a parallelogram is acted upon by two couples, as shown in the figure. Let us determine the angle β if the resultant couple is 1.8 N.m clockwise.

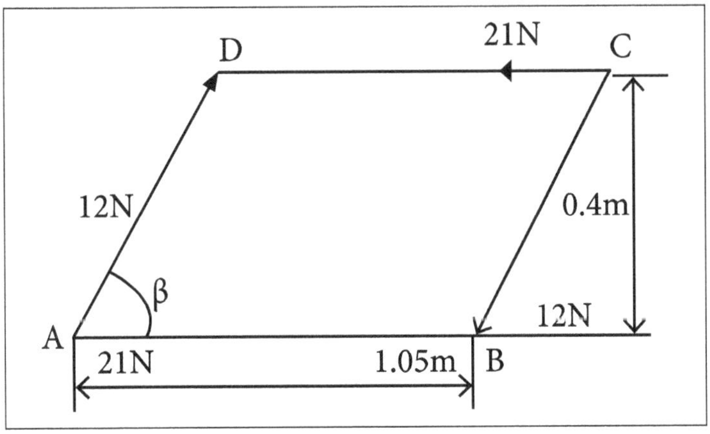

Solution:

Given:

In angle CBE,

$$\tan \beta = (0.4/BE)$$

$$BE = (0.4/\tan\beta)$$

$$AE = AB + BE$$

$$AE = \left(1.05\frac{0.4}{\tan\beta}\right)$$

Now taking moments about 'A' we get,

$$(21 \times 0.4) + (12\ \cos\beta \times 0.4) - (12\ \sin\beta \times AE)$$

It is given that resultant couple = 1.8 N.m

$$8.4 + 4.8\cos\beta - 12.6\sin\beta - 4.8\ \cos\beta = -1.8$$

$$12.6\ \sin\beta = 10.2$$

$$\sin\beta = (10.2/12.6)$$

$$\therefore \beta = 54^\circ$$

2.2 Equilibrium of Forces

Equilibrium is defined as the status of the body when it is subjected to a system of forces. We know that for a system of forces acting on a body the resultant can be found out, by Newton's 2rd Law of Motion. The body should then move in the direction of the resultant with some amount of acceleration. If the resultant force is equal to zero it implies that the net effect of the system of forces is equal to zero, this represents the state of equilibrium. For a system of coplanar concurrent forces for the resultant to be zero the equation is as follows,

$$\sum f_{x_1} = 0$$
$$\sum f_{y_i} = 0$$

2.2.1 Definition of Equilibrant

Equilibrant is a single force which when added to a system of forces brings the status of equilibrium. Hence this force is of the same magnitude as the resultant but opposite in sense.

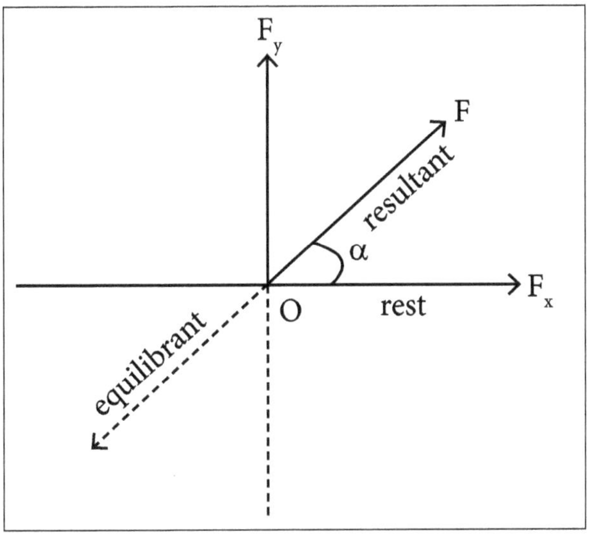

Equilibrant.

2.2.2 Conditions of Static Equilibrium for Different Force Systems

Conditions for equilibrium of a system of parallel forces acting in a plane:

- The algebraic sum of forces in zero. (i.e.) $\sum F = 0$.
- The algebraic sum of moments about any point is zero. (i.e.) $\sum M = 0$.

Equilibrium of Rigid Bodies in Two Dimensions

Necessary and sufficient conditions:

- $\sum F_X = 0$
- $\sum F_Y = 0$
- $\sum F_z = 0$
- $\sum M_X = 0$
- $\sum M_Y = 0$

A rigid body requires both the balance of force and moments to be in equilibrium while the balance of force prevents the body from translating and the balance of moments prevent it from rotating. As such the condition of equilibrium of a rigid body in dimensions will be as follows,

$$\sum FX = 0; \sum FY = 0 \text{ and } \sum M = 0$$

Where,

$\sum F_X$ = Algebraic sum of horizontal components of all the forces.

$\sum F_Y$ = Algebraic sum of vertical components of all the forces.

$\sum M$ = Algebraic sum of moments of all the forces acting on the body.

Equilibrium of Rigid Bodies in Three Dimensions

We have to consider the components in third dimension or z direction.

There are two methods for finding the unknown forces and moments:

- Vector equation of equilibrium.
- Scalar equations of equilibrium.

Example: Let us draw the free-body diagram of the shaft shown in the figure.

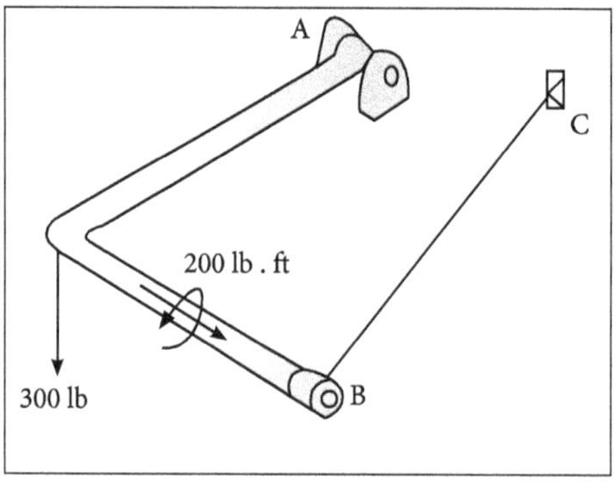

Pin at A and cable BC.

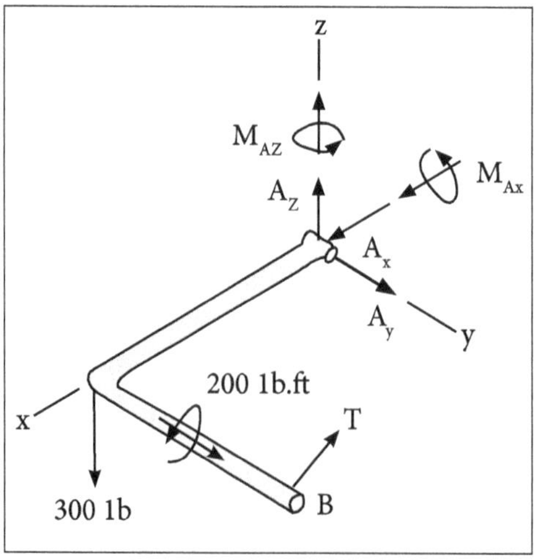

Consider the plane passing from points A and B as x-y plane.

Moment components are developed by the pin on the rod to prevent rotation about the x and z axes.

The string BC is a 2-force member which is under tension. Consider the tension T at point B.

In order to have a stable and proper system, the number of unknowns should be equal to that of equations. For example, we have 6 unknowns and we also have 6 equations.

$$\sum F_x = 0,\ \sum F_y = 0,\ \sum F_z = 0,\ \sum M_x = 0, \sum M_y = 0,\ \sum M_z = 0.$$

Improper Constraints

Number of equations can be equal to that of unknowns but the body could be unstable because of improper constraining by supports.

In 2-dimension problems the support reactions all intersect at the same point, so the constraints are improper. Here, the 100N force causes a moment which is not restrained by any of the support constraints.

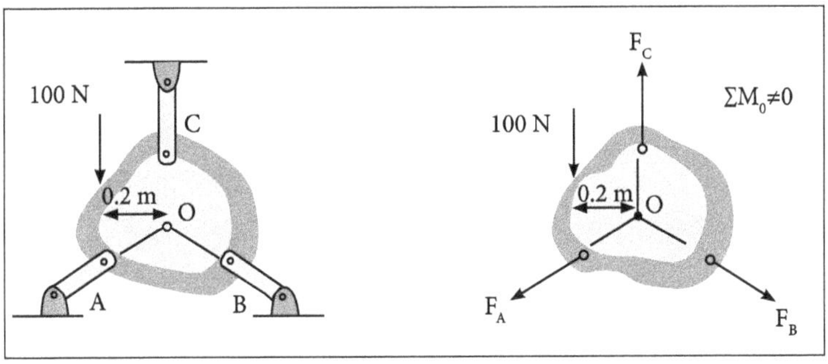

In 3-dimension problems like this, the support reactions all meet or intersect at a common axis, so the constraints are improper. Constraints A and B are not proper, since they do not resist the moment caused by 400N force.

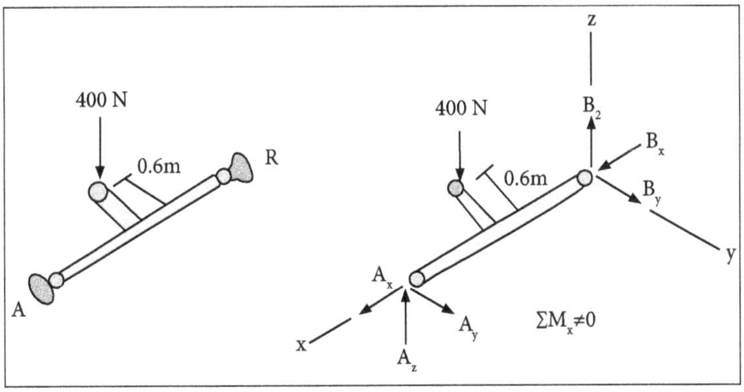

In the following figure, the reaction forces are all parallel, hence the constraints are improper. The body will now move under the exerted 100N force.

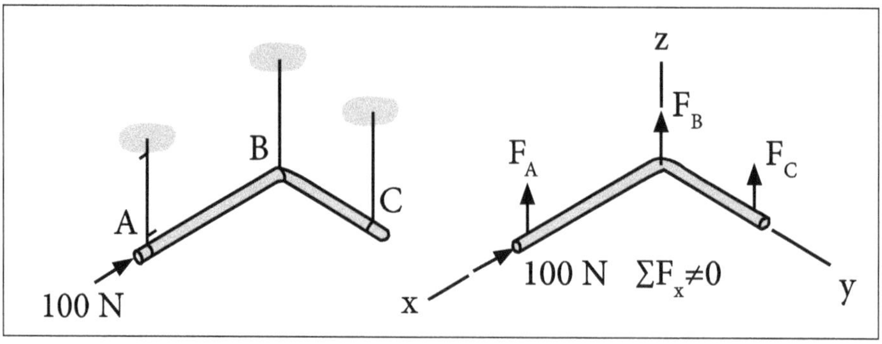

Redundant Constraints

Redundant supports are those, under which the body becomes statically indeterminate. Also, unknowns are more than equations. The below figures are the examples of redundant constraints.

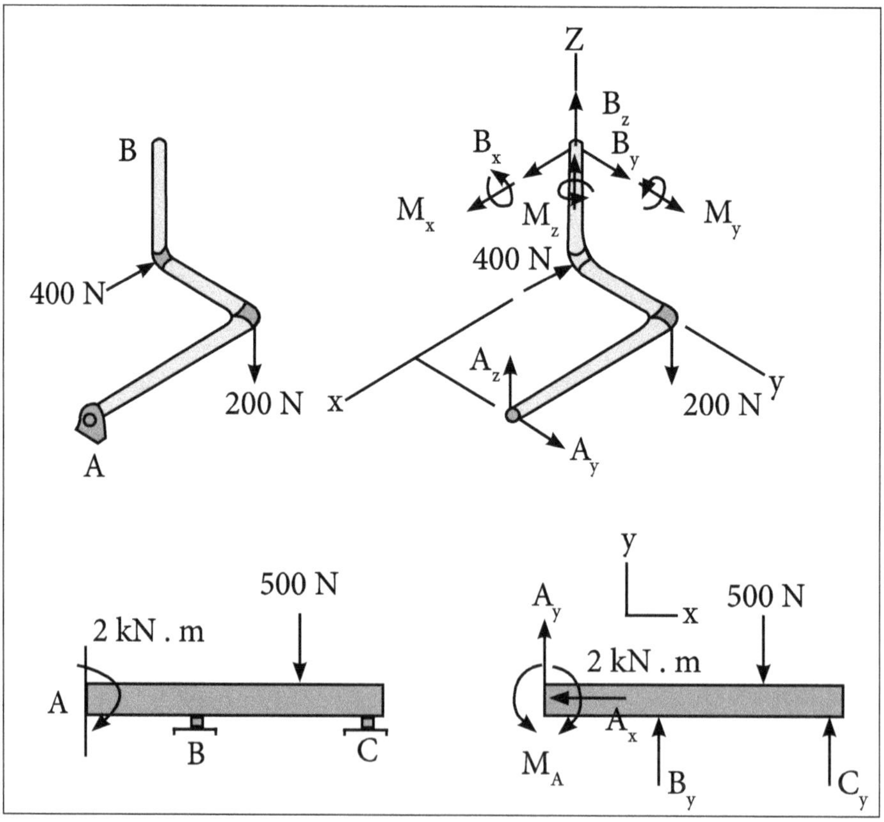

When there are fewer reactive forces than the equations of equilibrium, the constraints are called partially constrained.

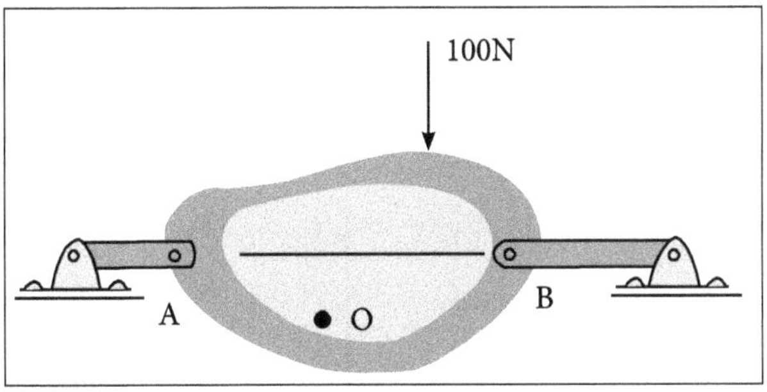

2.2.3 Lami's Theorem

It states that, "If the forces acting at a point are in equilibrium, each force will be proportional to the sine of the angle between the other two forces." Let the three forces P, Q and Rare acting at a point O and are in equilibrium as shown in below figure:

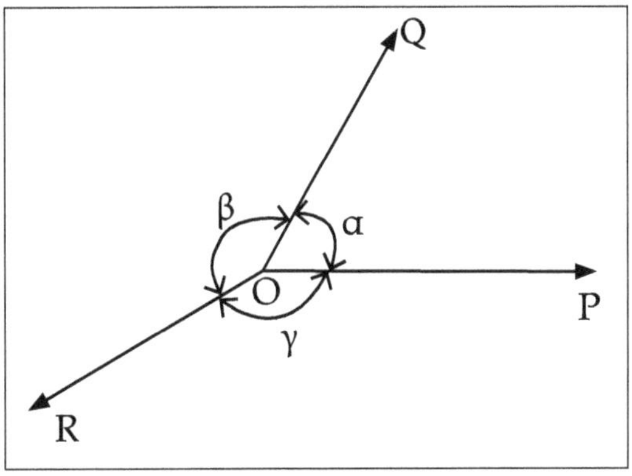

Let,

α = Angle between force P and Q.

β =Angle between force Q and R.

y = Angle between force R and P.

Then according to Lame's theorem, P is proportional sin of angle between Q and R, a sin β . P / sin β = constant.

Similarly Q / sin γ = constant, R / sin α = constant

P/sin β = Q/sin γ = R/sin α

Problems

1. In a jib crane, the jib and the tie rod are 5 m and 4 m long respectively. The height of crane post in 3 m and the ties rod remains horizontal. Let us calculate the forces produced in jib and tie rod when a load of 2 kN in suspended at the crane head.

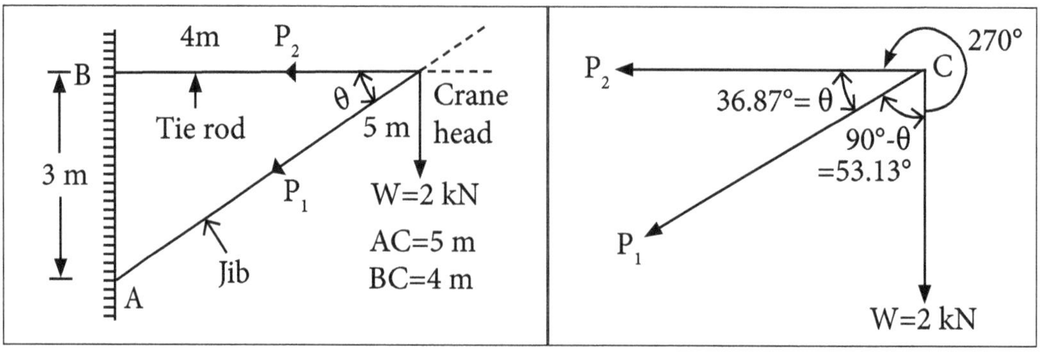

Solution:

Given:

Jib length = 5m

Tie rod length = 4m

Height of crane post = 3m

Suspended load = 2kN

To find:

Force produced in jib and tie rod

From figure,

$\sin q = 3/5 = 0.6$

$q = 36.87^{o}$

Let P_1 and P_2 be the forces developed in jib and tie rod respectively. The three forces P_1, P_2 and W are shown in figure with the angle between the forces calculated from the given directions.

The line of action of forces P_1, P_2 and weigh W meet at the point C and therefore Lami's theorem is applicable. That gives:

$$P_1/\sin 270° = P_2/\sin\ 53.13^{\circ} = 2/\sin\ 36.87^{\circ}$$

$$\therefore\ P_1 = 2\times\sin\ 270^{\circ}/\sin 36.87^{\circ} = 2\times 1/0. = -3.33\ \text{kN}$$

$$P_2 = 2\times\sin\ 53.13^{\circ}/\sin 36.87^{\circ} = 2\times 0.8/0.6 = 2.667\ \text{kN}$$

The -ve sign indicates that the direction of force P1 is opposite to that shown in figure. Obviously, the tie rod will be under tension and jib will in compression.

2. Let us resolve the 100 N force acting 30° horizontal into two components. One along horizontal and other along 120° to horizontal.

Solution:

Given:

Force = 100 N

The given force are:

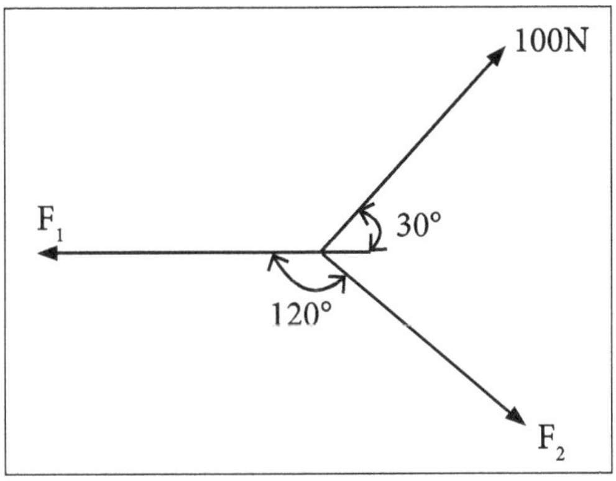

From the force diagram,

The angle between   and 100 N force is 90°.

The angle between F_1 and 100 N force is 150°.

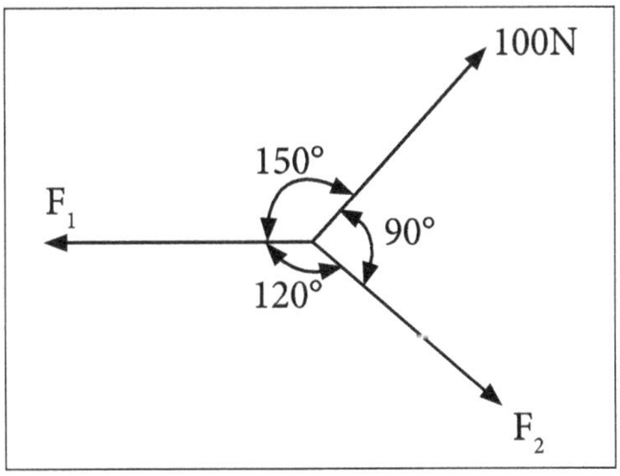

Applying Lami's Equation

$$\frac{F_1}{\sin 90^\circ} = \frac{100}{\sin 120^\circ} = \frac{F_2}{\sin 150^\circ}$$

$$F_1 = \frac{100 \times \sin 90}{\sin 120^\circ} = 115.47 \text{ N}$$

$$F_2 = \frac{100 \times \sin 150^\circ}{\sin 120^\circ} = 57.73 \text{ N}$$

3. Here we consider that a road roller of weight 5000 N, which is of cylindrical shape, is pulled by a force F at an angle of 30° with horizontal as shown in figure. It has to cross an obstacle of height 3 cm. Let us calculate the force F required to just cross this obstacle. The radius of the roller is equal to 30 cm.

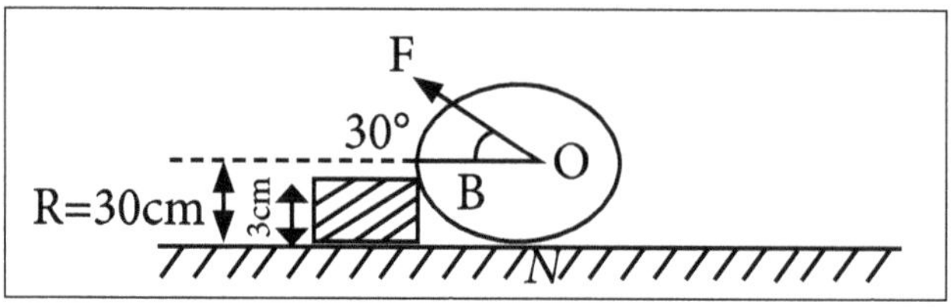

Solution:

Given:

Obstacle Height = 30mm.

Weight of Roller = 5000 N.

Radius of Roller = 600 mm.

Angle = 30°.

To find:

Force:

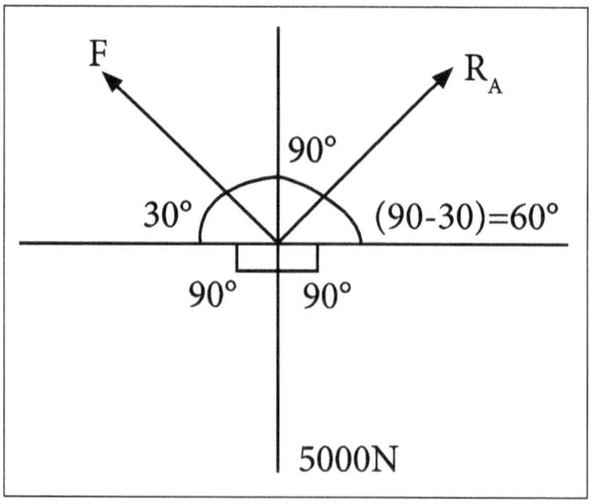

Free Body Diagram.

$$\frac{F}{\sin 150} = \frac{R_A}{\sin 120} = \frac{5000}{\sin 90^\circ}$$

$$F = \frac{\sin 150^\circ}{\sin 90^\circ} \times 5000$$

$$F = 2500 \text{ N}$$

$$R_A = \frac{\sin 120^\circ}{\sin 90^\circ} \times 5000$$

$$R_A = \frac{0.866}{1} \times 5000$$

$$R_A = 4330 \text{ N}$$

From,

Δ AOC

OC = Radius - AB

OC = 300 – 30

OC = 270 mm

$$\cos\theta = \frac{OC}{OA} = \frac{270}{300}$$

OA = Radius = 300 mm

$\theta = 30^\circ$.

Force F required to just chess F = 2500N.

4. In the figure shown, the three wires are joined at D.

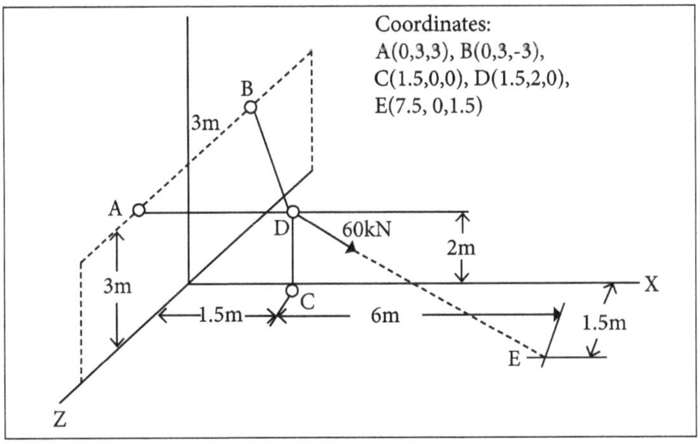

Two ends A and B are on the wall and the other end C is on the ground. The wire CD is vertical. A force of 60 kN is applied at 'D' and it passes through a point E on the ground as shown in figure. Let us determine the forces in all the three wires.

Solution:

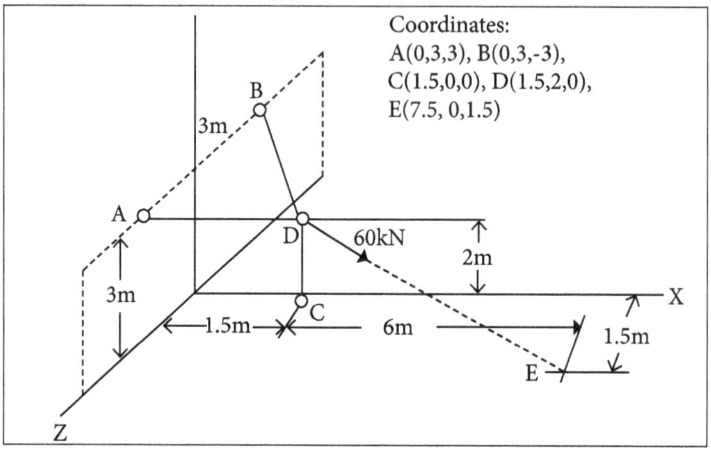

Given:

Let,

T_{AD} – Force acting in cable AD

T_{BD} – Force acting in cable BD

T_{CD} – Force acting in cable DC.

We know that,

$$T_{DA} = T_{DA} \times \lambda_{DA}$$

$$= T_{DA} \times \left[\frac{\left(\overrightarrow{OD} - \overrightarrow{OA}\right)}{\sqrt{(OA - OA)^2}} \right]$$

$$= T_{DA} \times \left[\frac{(1.5-0)i + (2-3)j + (0-3)k}{\sqrt{(1.5)^2 + (-1)^2 + (-3)^2}} \right]$$

$$= T_{DA} \times \left[\frac{1.5i - ij - 3k}{3.5} \right]$$

$$= T_{DA} \times (1.5i - 0.285j - 0.857k) \quad ...(1)$$

$$T_{DB} = T_{DB} \times \lambda_{DB}$$

$$= T_{DB} \times \left[\frac{(1.5-0)i+(2-3)j+(0-(-3))k}{\sqrt{(1.5)^2+(-1)^2+(3)^2}} \right]$$

$$= T_{DB} \times \left[\frac{1.5i+(-1)j+3k}{3.5} \right]$$

$$T_{DB} = T_{DB} \times [0.428i-0.285j+0.8571k] \quad \ldots(2)$$

$$T_{DC} = T_{DC} \times \lambda_{DC}$$

$$= T_{DC} \times \left[\frac{(1.5-1.5)i+(2-0)j+(0-0)k}{\sqrt{D^2+2^2+0^2}} \right]$$

$$= T_{DC} \times \frac{0i+2j+k}{2}$$

$$T_{DC} = T_{DC} \times (0i+1j+0.5k) \quad \ldots(3)$$

Then force on cable DE is given by T_{DE},

$$T_{DE} = T_{DE} \times \left[\frac{(1.5-7.5)i+(2-0)j}{1} \right] T$$

$$= T_{DE} \times \left[\frac{(1.5-7.5)i+(2-0)j+(0-1.5)k}{\sqrt{6^2+2^2+(-1.5)^2}} \right]$$

$$= T_{DE} \times \left[\frac{-6i+2j-1.5k}{6.5} \right]$$

$$T_{DE} = T_{DE} \times (-0.923i+0.3076j-0.23k) \quad \ldots(4)$$

We know that,

$T_{DE} = 60$ kN is gives in the problem,

$$T_{DE} = (-0.923i+0.3076j-0.23k)$$

$$T_{DE} = \left(-\ 55.38i + 18.45j - 13.8k\right).$$

Algebraic sum of all forces is zero,

$$\sum F_X = 0$$

$$\sum F_Y = 0$$

$$\sum F_Z = 0$$

$\sum F_X = 0$ from equation (1), (2), (3), (4)

$$\sum 1.5\ T_{DA} + 0.428\ T_{DB} + 0\ T_{DC} - 55 = 0 \qquad ...(5)$$

$\sum F_Y = 0$ from equation (1), (2), (3), (4)

$$-\ 0.285\ T_{DA} - 0.285\ T_{DB} + 1\ T_{DC} = 18.45 = 0 \qquad ...(6)$$

$$\sum F_2 = 0$$

$$-\ 0.857\ T_{DA} + 0.857\ T_{DB} + 0.5\ T_{DC} + \left(-\ 13.8\right) = 0 \qquad ...(7)$$

From equation (5),

$$1.5\ T_{DA} + 0.428\ T_{DB} = 55$$

$$T_{DA} = \left(\frac{55 - 0.428\ T_{DB}}{1.5}\right)$$

$$T_{DA} = 36 - 0.285\ T_{DB}$$

Substitute T_{DA} in equation (6),

$$-\ 2.285\ \left(36 - 0.285\right)T_{DB} - 0.255\ T_{DB} + \ T_{DC} + \ 16.45 = 0$$

$$\left(-\ 10.28 + 0.08125\right)\ T_{DB} - \ T_{DB} \times 0.285 + T_{DC} + 18.45 = 0$$

$$\left(-\ 10.17\right)\ T_{DB} - T_{DB} \times\ 0.285 + T_{DC} + 18.45 = 0$$

$$-\ 10.46\ T_{DB} + \ T_{DC} + 18.45 = 0$$

$$T_{DC} = 18.45 + 10.46\ T_{DB}.$$

Substitute above equation in equation (7),

$$-0.857(36-0.285\,T_{DB})+0.857\,(T_{DB})+0.5(18.45+10.46\,T_{DB})-13.8=0$$

$$-30.852+0.244\,T_{DB}+0.857\,T_{DB}+9.225+5.23\,T_{DB}-13.8=0$$

$$6.331\,T_{DB}-34.575=0$$

$$T_{DB}=5.461\text{ kN}$$

From,

$$T_{DA}=36-0.285\,T_{DB}$$

$$=36-0.285\,(5.461)$$

$$=36-1.556$$

$$T_{DA}=34.44\text{ kN}$$

From,

$$T_{DC}=18.45+10.46\,T_{DB}$$

$$=18.45+10.46\,(5.461)$$

$$T_{DC}=75.57\text{ kN.}$$

Result:

Forces in All the cables are listed below,

$$T_{AD}=34\text{kN}$$

$$T_{BD}=5\text{kN}$$

$$T_{CD}=75\text{kN}$$

5. Let the block P = 5 kg and block Q of mass m kg is suspended through the chord is in the equilibrium position as shown in the figure. Let us determine the mass of block Q.

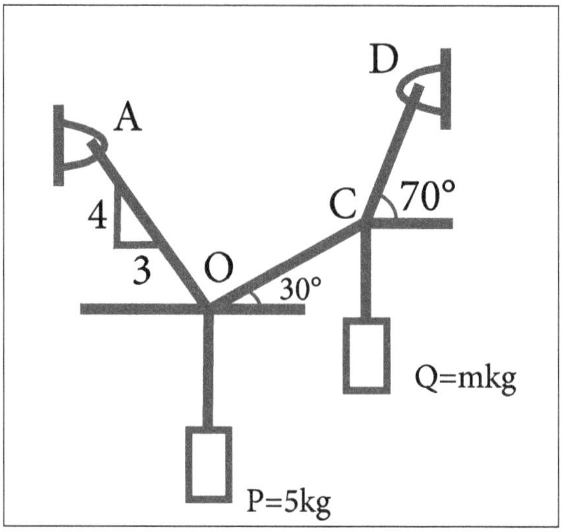

Solution:

Given:

Block P = 5 kg

To find the Inclination of chord AB,

$\tan\theta = 4/3$

$\tan\theta = 1.333$

$\theta = 53.13°$.

The forces acting in the chord along with the weights attached at B and C are shown in following figure:

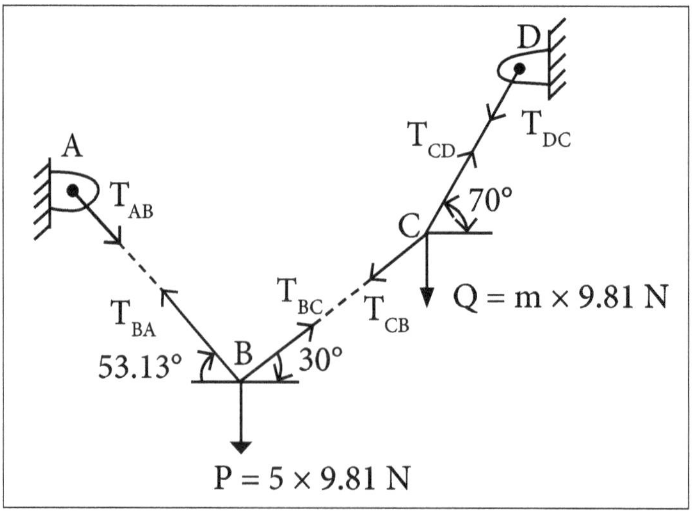

Apply Lami's Equation at B,

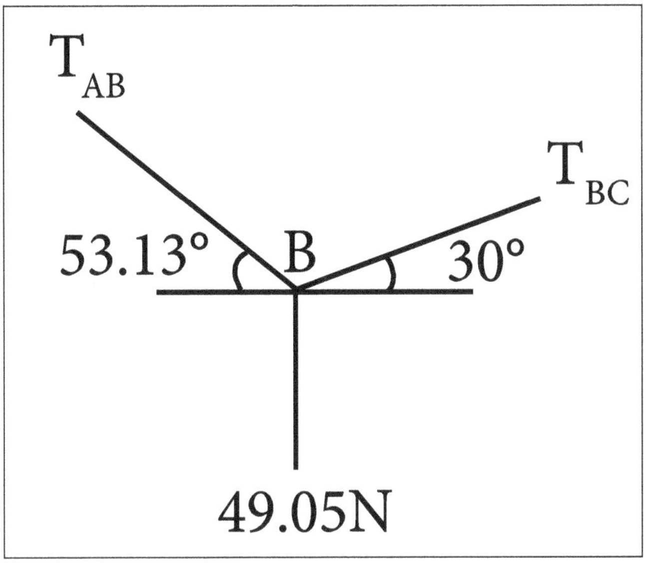

From the geometry of figure, angle between T_{AB} and T_{BC} is,

$$\left[180-(53.13+30)\right]=96.87^{\circ}$$

Angle between T_{BC} and 49.05 N is (90° + 30°) = 120°

Angle between 49.05 N and T_{AB} is (90 + 53.13) = 143.13°.

$$\frac{T_{AB}}{\sin 120^{\circ}}=\frac{T_{BC}}{\sin(90^{\circ}+53.13^{\circ})}=\frac{49.05}{\sin\left[180-(53.13+30)\right]}$$

$$\frac{T_{AB}}{\sin 120^{\circ}}=\frac{T_{BC}}{\sin 143.13^{\circ}}=\frac{49.05}{\sin 96.87}$$

$$T_{AB}=\frac{49.05\times\sin 120}{\sin 96.87}$$

$$T_{AB}=42.786\text{ N}$$

$$T_{BC}=\frac{49.05\times\sin 143.13^{\circ}}{\sin 96.87^{\circ}}$$

$$T_{BC}=29.65\text{ N}$$

Apply Lami's Equation at C,

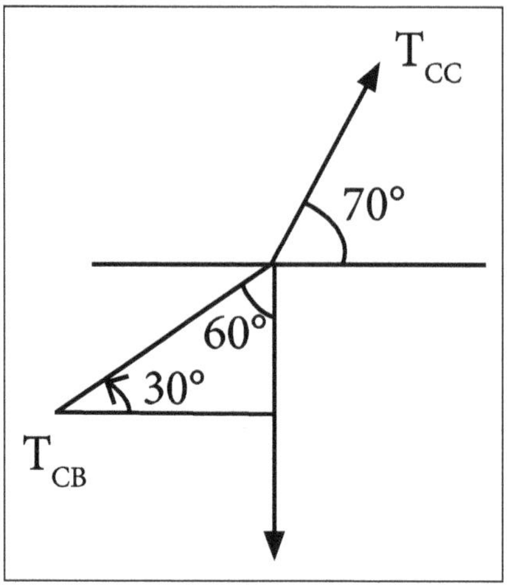

$$Q = m \times 9.81 \text{ N}$$

From the geometry of figure, angle between T_{CB} and Q is 60°

Angle between T_{CD} and Q is (90° + 70°) = 160°;

Angle between T_{CD} and T_{CB} is 360 – (60 + 90 + 70) = 140°

$$\frac{T_{CB}}{\sin 160^\circ} = \frac{T_{CD}}{\sin 60^\circ} = \frac{m \times 9.81}{\sin 140^\circ}$$

Where,

$$T_{CB} = T_{BC} = 29.65 \text{ N}$$

$$\frac{29.65}{\sin 160^\circ} = \frac{T_{CD}}{\sin 60^\circ} = \frac{m \times 9.81}{\sin 140^\circ}$$

$$T_{CD} = \frac{29.65 \times \sin 60^\circ}{\sin 160^\circ}$$

$$T_{CD} = 75.08 \text{ N}$$

$$m \times 9.81 = \frac{29.65 \times \sin 140^\circ}{\sin 160^\circ}$$

$$m \times 9.81 = 55.72$$

$$m = 5.68 \text{ kg}$$

Result:

$T_{AB} = 42.786N$

$T_{BC} = 29.65\ N$

$T_{CD} = 75.08\ N$

m = 5.68 kg (mass of the block Q).

6. Let us determine the tensions in various segments of the connected flexible cables as shown in the figure.

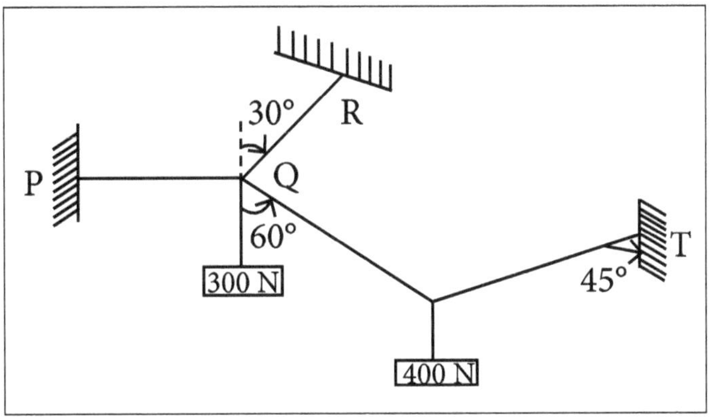

Solution:

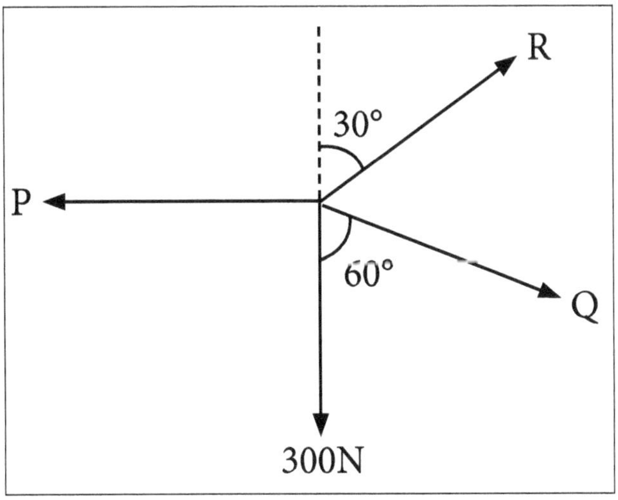

$$\sum V = 0$$

$$-300 + R\cos 30° - Q\cos 60° = 0 \quad ...(1)$$

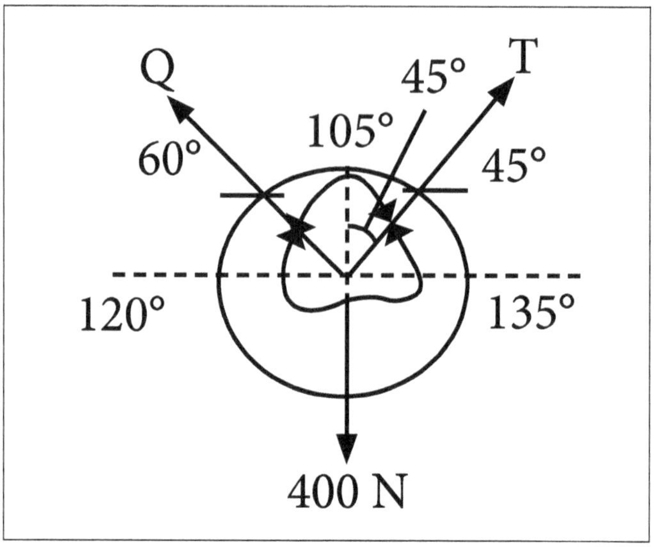

$$\frac{400}{\sin 105^\circ} = \frac{T}{\sin 120^\circ} = \frac{Q}{\sin 135^\circ}$$

By Lami's Theorem

$T = 358.63$ N

$Q = 414.110 \times \sin 135^\circ$

$Q = 292.89$ N

From equation (1),

$R = (446.41/\cos 30^\circ)$

$R = 515.46$ N

$\sum H = 0$

$-P + (R \times \sin 30^\circ) + (292.82 \times \sin 60^\circ) = 0$

$P = 511.32$ N

7. Let us consider the 75-kg crate shown in the space diagram of figure. This crate was lying between two buildings and it is now being lifted onto a truck, which will remove it. The crate is supported by a vertical cable, which is joined at A to two ropes which pass over pulleys attached to the buildings at B and C. Let us determine the tension in each of the ropes AB and AC.

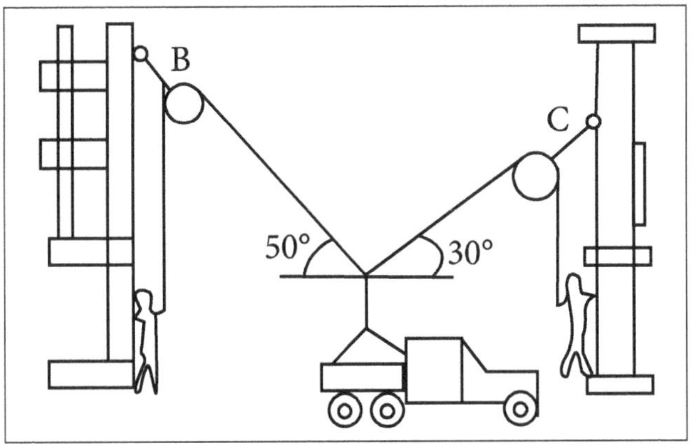

Solution:

Given:

Crate = 75-kg

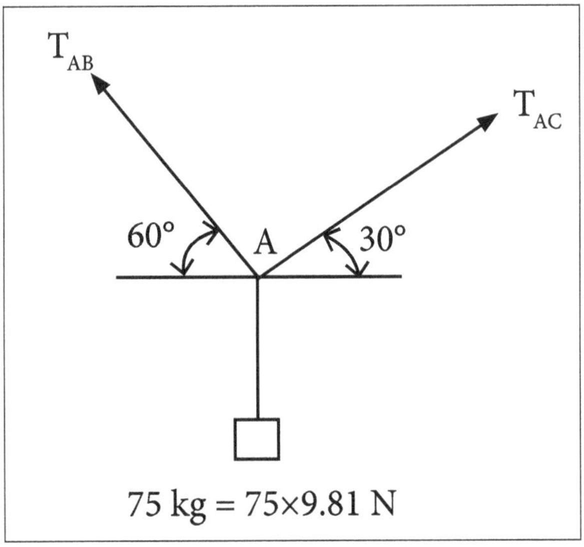

Free body diagram at A,

T_{AB} = Tension in the rope AB, from A to B

T_{AC} = Tension in the rope AC, from A to C.

From the geometry of the figure, the angle between T_{AC} and 75 kg is (30° + 90°) = 120°

Angle between T_{AB} and 75 kg is (50° + 90°) = 140°;

Angle between T_{AB} and T_{AC} is (180° - 50° - 30°) = 100°.

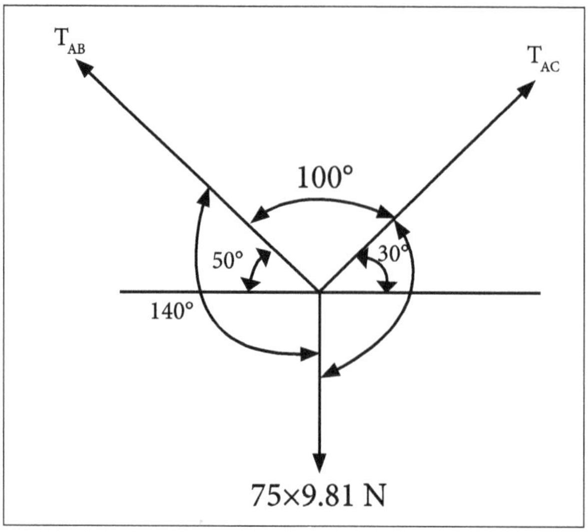

Applying Lami's equation at A,

$$\frac{T_{AB}}{\sin 120°} = \frac{T_{AC}}{\sin 140°} = \frac{75 \times 9.81}{\sin 100°} \quad ...(1)$$

From Equation 1,

$$\frac{T_{AB}}{\sin 120°} = \frac{75 \times 9.81}{\sin 100°}$$

$$T_{AB} = \frac{\sin 120° \times 75 \times 9.81}{\sin 100°}$$

$$T_{AB} = 647\ \text{N}$$

From Equation 1,

$$\frac{T_{AC}}{\sin 140} = \frac{75 \times 9.81}{\sin 100}$$

$$T_{AC} = \frac{\sin 140 \times 75 \times 9.81}{\sin 100}$$

$$T_{AC} = 48023\ \text{N}$$

Result:

$$T_{AB} = 647\,\text{N}$$

$$T_{AC} = 480.23 \text{ N}$$

Numeric Problems

1. Let us determine the magnitude and direction of the resultant of the coplanar concurrent force system shown in figure below.

Solution:

Given:

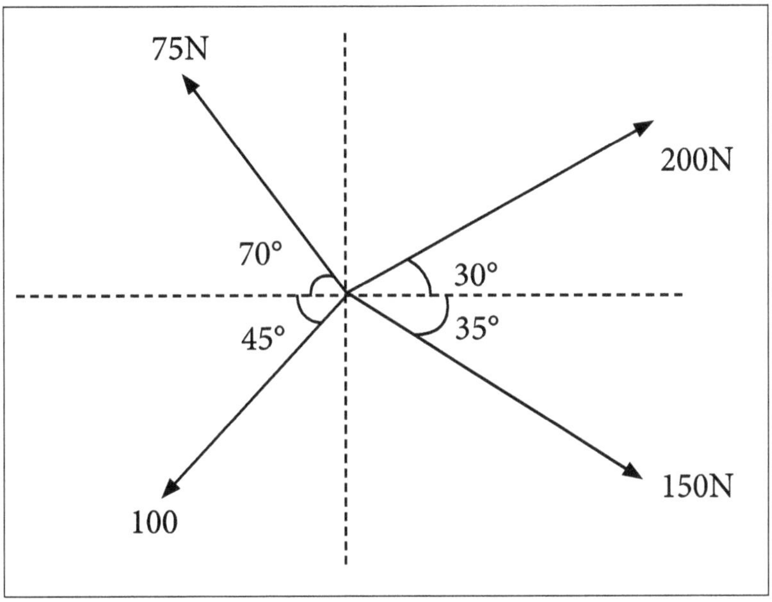

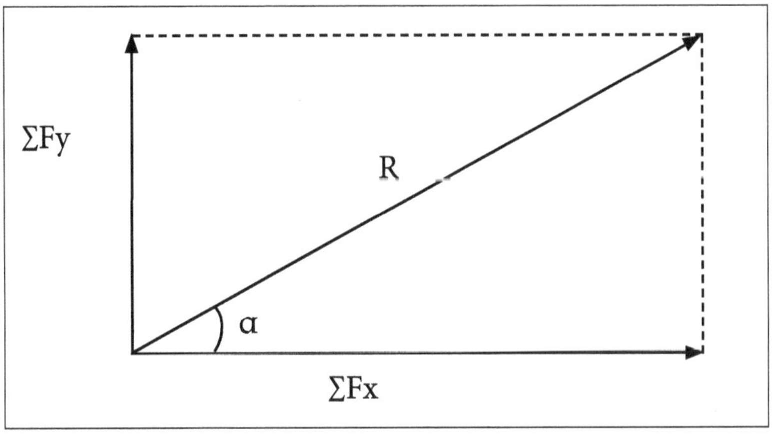

To find:

Magnitude and direction of the resultant of the coplanar concurrent force system,

Let R be the given resultant force system.

Let α be the angle made by the resultant with x- direction.

The magnitude of the resultant is given as,

$$R = \sqrt{(\sum Fx)^2 + (\sum Fy)^2}$$

$$\alpha = \tan^{-1}\left(\frac{\sum Fy}{\sum Fx}\right)$$

$$\sum F_x = 200 \cos 30° - 75 \cos 70° - 100 \cos 45° + 150 \cos 35°$$

$$\sum F_x = 199.7 \text{ N}$$

$$\sum F_y = 200 \sin 30° + 75 \sin 70° - 100 \sin 45° - 150 \sin 35°$$

$$\sum F_y = 13.72 \text{ N}$$

$$R = \sqrt{(\sum Fx)^2 + (\sum Fy)^2}$$

$$R = 200.21 \text{ N}$$

$$\alpha = \tan^{-1}\left(\frac{\sum Fy}{\sum Fx}\right)$$

$$\alpha = \tan\ (13.72/199.72) = 3.93°$$

2. Let us determine the resultant of the concurrent force system shown in figure.

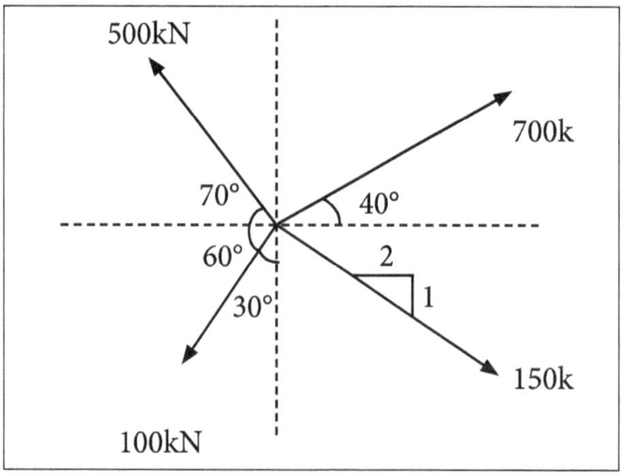

Solution:

Given:

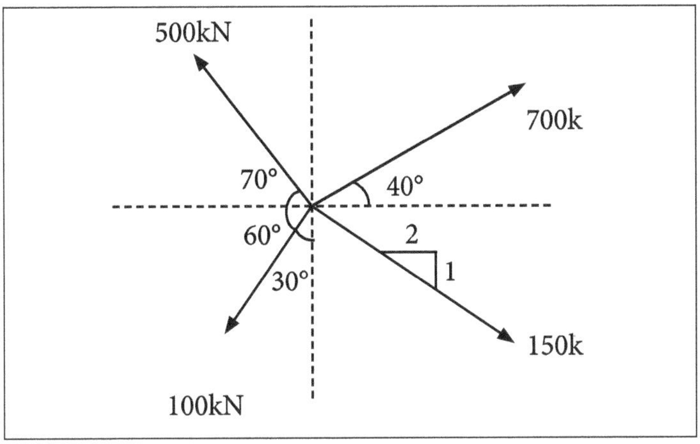

To find:

Resultant of concurrent force system

Let R be the given resultant force system.

Let α be the angle made by the resultant with x- direction.

The magnitude of the resultant is given as,

$$R = \sqrt{\left(\sum Fx\right)^2 + \left(\sum Fy\right)^2}$$

$$\alpha = \tan^{-1}\left(\frac{\sum Fy}{\sum Fx}\right)$$

$$\sum F_x = 700 \cos 40° - 500 \cos 70° - 800 \cos 60° + 200 \cos 26.56°$$

$$\sum F_x = 144.11 \text{ KN}$$

$$\sum F_y = 700 \sin 40° + 500 \sin 70° - 800 \sin 60° - 200 \sin 26.56°$$

$$\sum F_y = 137.55 \text{ KN}$$

$$R = \sqrt{\left(\sum Fx\right)^2 + \left(\sum Fy\right)^2}$$

$$R = 199.21 \text{ N}$$

$$\alpha = \tan^{-1}\left(\frac{\sum Fy}{\sum Fx}\right)$$

$$\alpha = \tan^{-1}(137.55/144.11) = 43.66°$$

3. Let us determine the resultant of a coplanar concurrent force system shown in figure below.

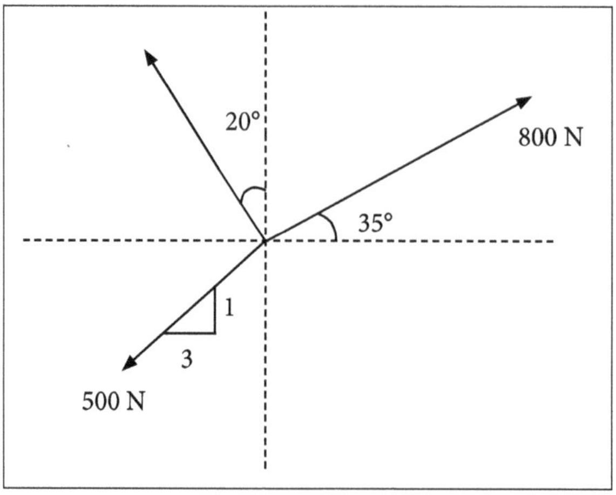

Solution:

Given:

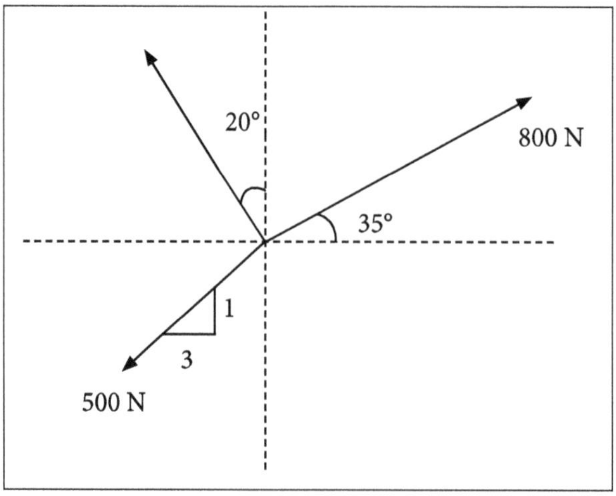

To find:

Resultant of coplanar concurrent force system

Let R be the given resultant force system.

Let α be the angle made by the resultant with x- direction.

The magnitude of the resultant is given as,

$$R = \sqrt{\left(\sum F_x\right)^2 + \left(\sum F_y\right)^2}$$

$$\alpha = \tan^{-1}\left(\frac{\sum Fy}{\sum Fx}\right)$$

$$\sum F_x = 800\ \cos\ 35^\circ - 100\cos 70^\circ + 500\cos 60^\circ + 0$$

$$\sum F_x = 1095.48\ N$$

$$\sum F_y = 800\sin\ 35^\circ + 100\sin\ 70^\circ + 500\sin\ 60^\circ - 600$$

$$\sum F_y = 110.90\ N$$

$$R = 1101.08\ N$$

$$\alpha = \tan^{-1}\left(\frac{\sum Fy}{\sum Fx}\right)$$

$$\alpha = \tan^{-1}(110.90/1095.48) = 5.78^\circ$$

4. Let us determine the magnitude and direction of the resultant of the coplanar concurrent force system shown in figure.

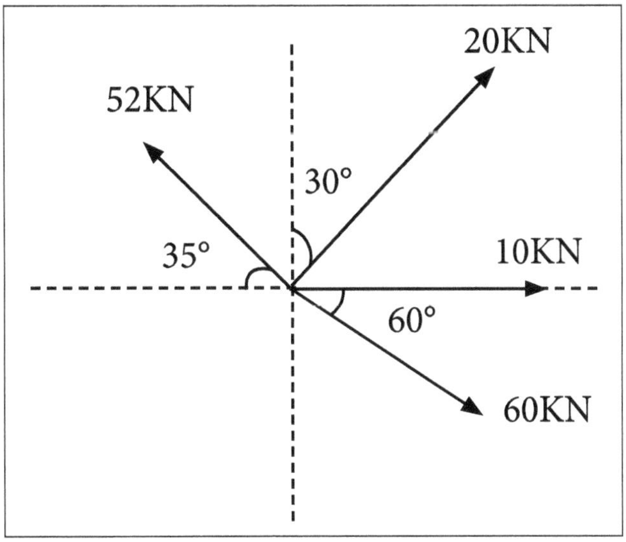

Solution:

Given:

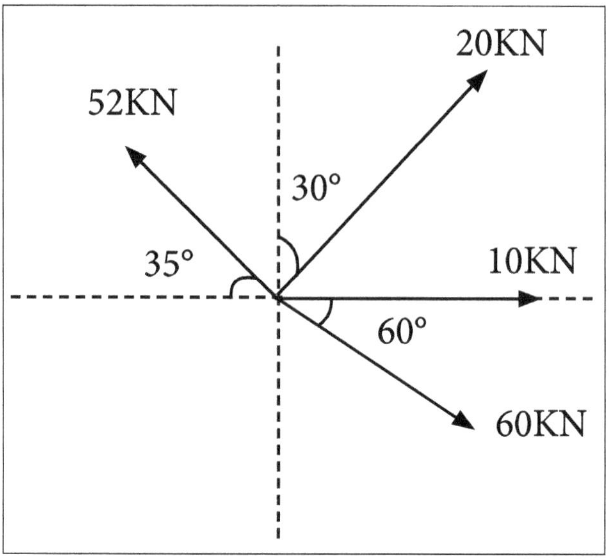

To find:

Magnitude and direction of the resultant of the coplanar concurrent force system

Let R be the given resultant force system.

Let α be the angle made by the resultant with x- direction.

The magnitude of the resultant is given as,

$$R = \sqrt{\left(\sum Fx\right)^2 + \left(\sum Fy\right)^2}$$

$$\alpha = \tan^{-1}\left(\frac{\sum Fy}{\sum Fx}\right)$$

$$\sum F_x = 20\ \cos\ 60° - 52\cos\ 30° + 60\cos 60° + 10$$

$$\sum F_x = 7.404\ \text{KN}$$

$$\sum F_y = 20\ \sin 60° + 52\sin\ 30° - 60\sin\ 60° + 0$$

$$\sum F_y = -8.641\ \text{KN}$$

$$R=\sqrt{\left(\sum Fx\right)^{2}+\left(\sum Fy\right)^{2}}$$

R = 11.379 N

$$\alpha=\tan^{-1}\frac{\left(\sum Fy\right)}{\left(\sum Fx\right)}$$

α = tan-1(−8.641/7.404) = −49.40°

2.3 Application of Static Friction in Rigid Bodies in Contact

Whenever a body moves or tends to move over another body or surface, a force which opposes the motion of the body is developed tangentially to the surface of contact, such an opposing force developed is called friction or frictional resistance.

The development of frictional resistance is due to the interlocking of the surface irregularities at the contact surface between two bodies. Consider a body weighing W resting on a rough plane and subjected to a force P to displace the body.

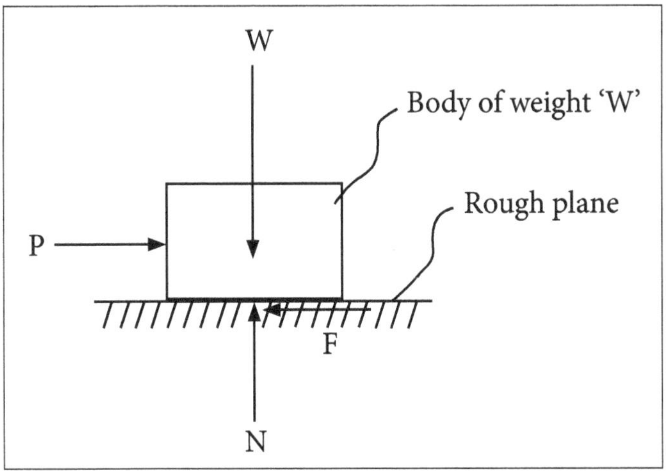

Where,

N = Normal reaction from rough surface

P = Applied force

F = Frictional resistance

W = Weight of the body

The body can start moving or sliding over the plane if the force P overcomes the frictional force F. The frictional resistance developed is directly proportional to the magnitude of the applied force which is responsible for causing motion up to a certain limit.

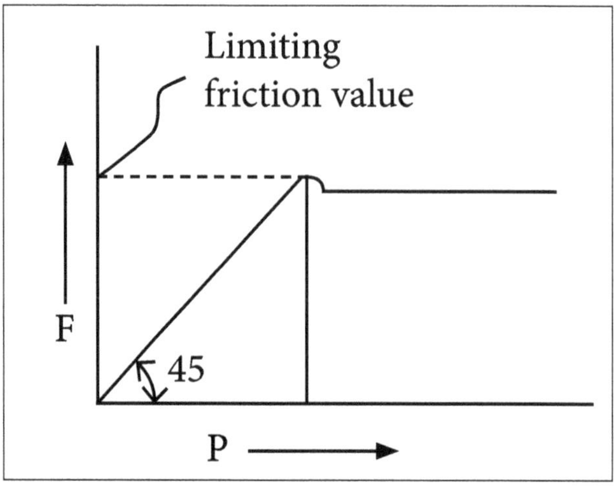

From the above graph we see that as P increases, F also increases. However F cannot increase beyond a particular limit. Beyond this limit the frictional resistance becomes constant for any value of the applied force.

If the magnitude of the applied force is less than the limiting friction value, the body is in equilibrium or remains at rest. If the magnitude of the applied force is greater than the limiting friction value the body starts moving over the surface.

When a body is moving it experiences a friction and this is known dynamic friction. The friction experienced by a body when it is in equilibrium or at rest is known as static friction. It can range between zero to a limiting fraction value.

The dynamic friction experienced by a body as it slides over a plane, as it is shown in figure is called sliding friction.

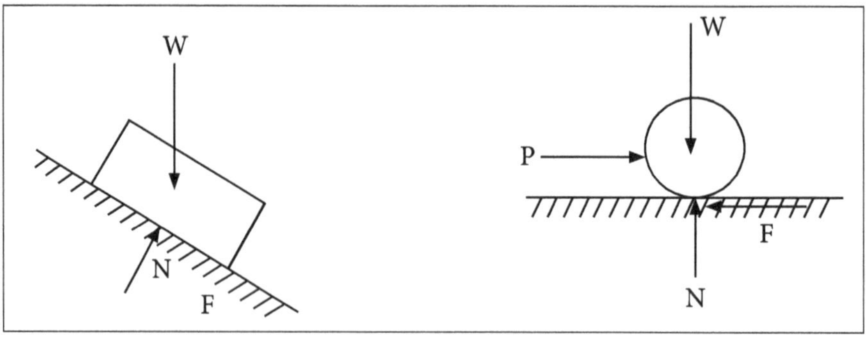

Sliding friction and Rolling friction.

The dynamic friction experienced by a body as it rolls over surface, as shown in figure is called rolling friction.

2.3.1 Types of Friction, Laws of Static Friction, Limiting Friction and Angle of Friction

Friction may be defined as a force of resistance acting on a body which prevents or retards slipping of the body relative to a second body and surface with which it is in control:

- Laws of static friction.
- Laws of Kinetic or dynamic friction.

Laws of Static Friction Following are the Laws of Static Friction

The force of friction always acts in a direction, opposite to that in which the body tends to move, if the force of friction would have been absent.

The magnitude of the force of friction is exactly equal to the force, which tends to move the body.

The magnitude of the limiting friction bears a constant ratio to the normal reaction between the two surfaces, Mathematically,

$F/R = \text{Constant}$

$F = \text{Limiting force}$

$R = \text{Normal force}$

The force of friction is independent of the area of contact between the two surfaces.

The force of friction depends upon the roughness of the surfaces.

Laws of Kinetic or Dynamic Friction Following are the Laws of Kinetic or Dynamic Friction

The force of friction always acts in a direction, opposite to that in which the body is moving.

The magnitude of kinetic friction bears a constant ratio to the normal reaction between the two surfaces. But the ratio is slightly less than that in case of limiting friction.

For moderate speeds, the force of friction remains constant. But it decreases slightly with the increase of speed.

Motion of Bodies

Wedge

A wedge is in general a triangular object that is placed between two objects to either hold them in place or is used to move one relative to the other. For example, the subsequent diagram shows a wedge under a block that is supported by the wall.

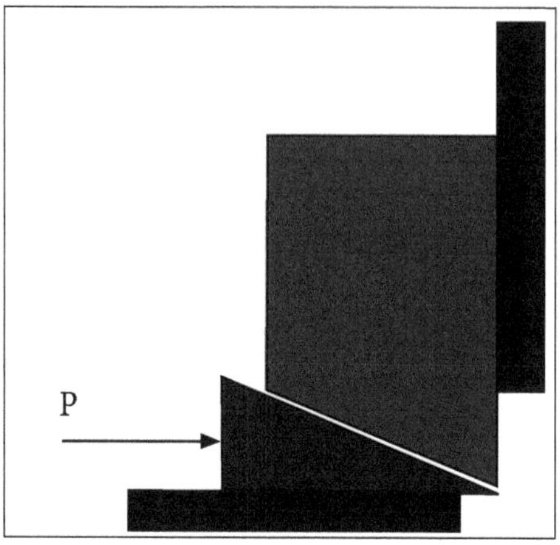

If the force P is large enough to push the wedge forward, then the block can rise and the following is an appropriate free-body diagram. Note that for the wedge to move one requires to have slip on all three surfaces. The direction of the friction force on every surface will oppose the slipping.

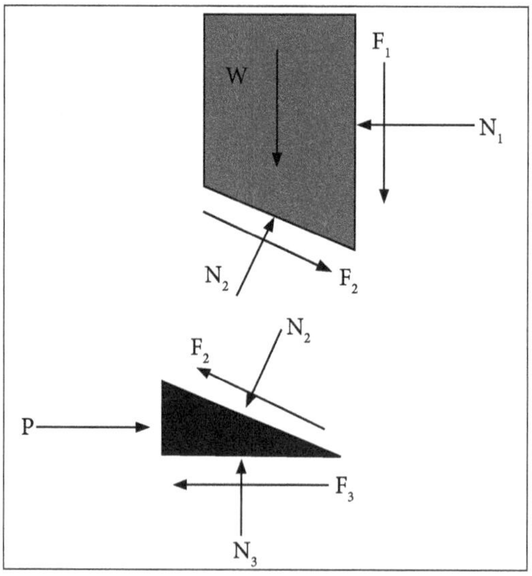

Since before the wedge will move each surface must overcome the resistance to slipping, one will assume that,

$$F_1 = \mu N_1$$

$$F_2 = \mu N_2$$

$$F_3 = \mu N_3$$

These equations and the equations of equilibrium are combined to solve the problem. If the force P is not large enough to hold the top block from coming down, then the wedge can be pushed to the left and the appropriate free-body diagram is the following:

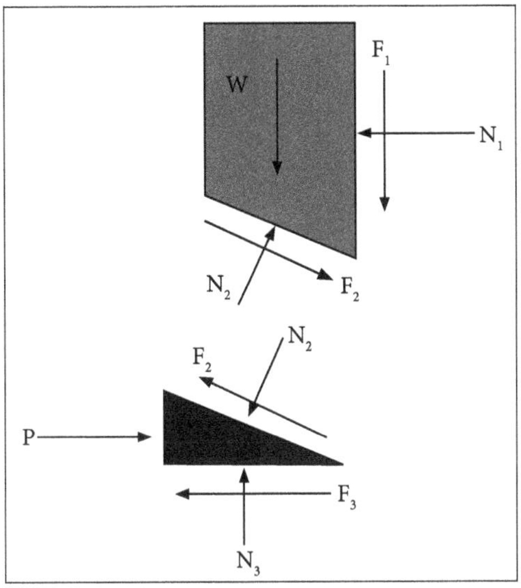

Ladder Friction

Consider a ladder AB resting on a rough horizontal surface and leaning against a rough wall. Then by its own weight or when a man starts climbing upon it, it tends to slide down.

At the point of impending motion, we know that the point A tends to slide downwards and the point B tends to slide away from the wall. Hence, apart from normal reactions at these points, frictional forces act in the directions as indicated trying to oppose its motion.

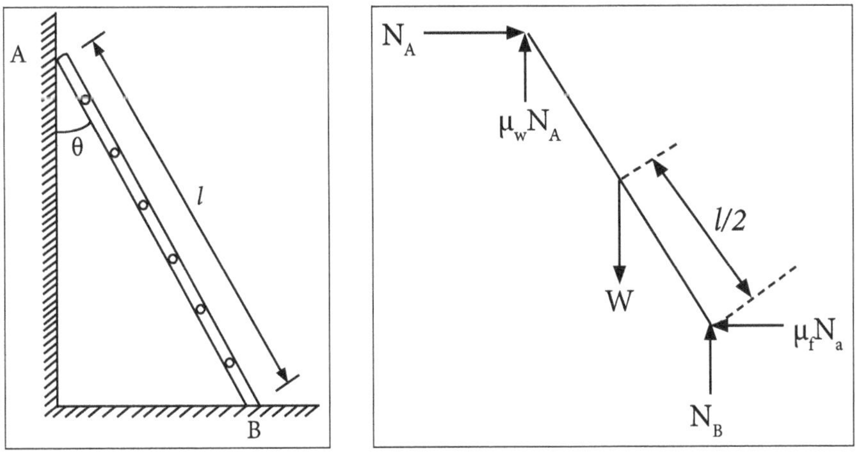

(a) (b)

In the figure (b) above, W is the weight of the ladder, μ_W and μ_f are the co-efficient of friction between the ladder and the wall and between the ladder and the floor respectively.

The force of friction at the wall acts upwards preventing the downward motion of the end A and its magnitude is equal to $\mu_W\ N_A$ at the point of impending motion.

Similarly, the force of friction at the floor acts towards the wall preventing the motion of the end B away from the wall and its magnitude is equal to $\mu_f\ N_B$ at the point of impending motion.

If a man climbs up the ladder, his weight must also be shown in the free-body diagram. At the point of impending motion, we can apply the conditions of equilibrium.

$$\Sigma F_x = 0$$

$$\Sigma F_y = 0$$

$$\Sigma M = 0$$

Belt Friction

Consider a flat belt passing over a fixed cylindrical drum, the motion between the belt and the drum is assumed to be impending.

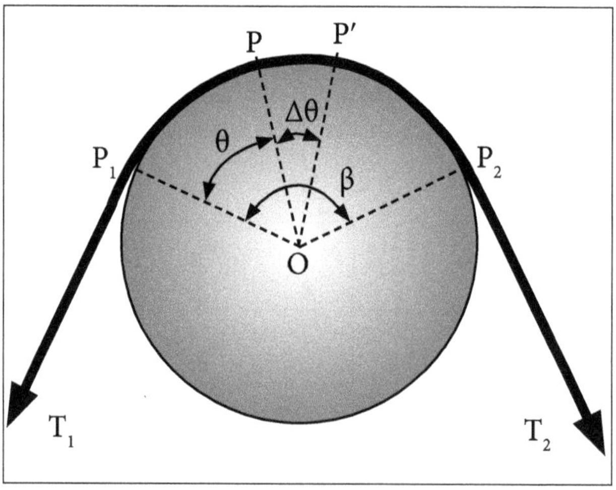

Consider a small element PP’ subtending an angle Δθ, the equation of equilibrium are,

$$\sum F_x = 0 (T + \Delta T)\ \cos\ (\Delta\theta/2) - T\cos(\Delta\theta/2) - \mu_s\ \Delta N = 0$$

$$\sum F_x = 0\ \Delta N - (T + \Delta T)\sin(\Delta\theta/2) - T\sin(\Delta\theta/2) = 0$$

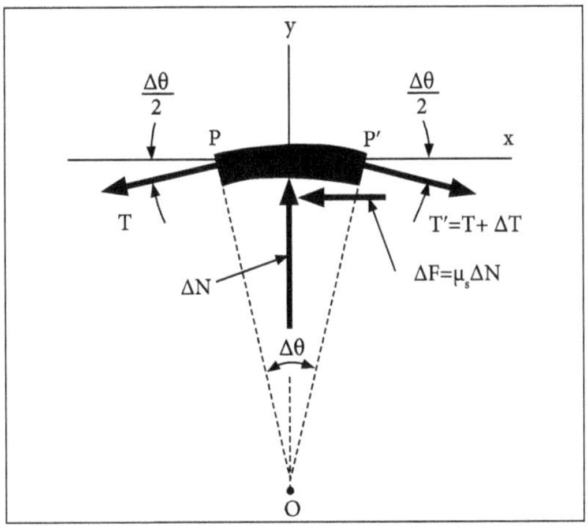

$$\Delta T\ \cos\ (\Delta\theta/2) - \mu_s(2T + \Delta T)\sin(\Delta\theta/2) = 0$$

or,

$$\frac{\Delta T}{\Delta\theta}\cos\Delta\theta/2 - \mu_s(T + \Delta T/2)\frac{\sin\Delta\theta/2}{\Delta\theta/2} = 0$$

For $\Delta\theta \to 0,\ \Delta N, \Delta F, \Delta T \to 0$

And $\cos\ \Delta\theta/2 \to 1$

$$\lim_{\Delta\theta\to 0}\frac{\sin\Delta\theta/2}{\Delta\theta/2} = \lim_{\Delta\theta\to 0}\frac{\frac{1}{2}\cos\Delta\theta/2}{\frac{1}{2}} = 1$$

And the equation becomes,

$$dT/d\theta - \mu_s T = 0$$

Integrating from P_1 to P_2,

$$\int_{T_1}^{T_2}\frac{dT}{T} = \mu_s\int_0^{\beta} d\theta$$

At $P_1:\ T = T_1,\ \theta = 0$

At $P_2:\ T = T_2,\ \theta = \beta$

It is obtained,

$$\ln\ T_2 - \ln T_1 = \mu_s\beta$$

$$\ln\ T_2/T_1 = \mu_s\beta$$

Or, $T_2 / T_1 = e_s^{\mu} \beta$

T_2 is always larger then T_1, T_2 therefore represents the tension in that part of the belt or rope which pulls, while T_1 is the tension in the part which resists. Angle of contact β may be expressed in radian and may be larger than 2 if the belt is actually slipping, s should be changed to K.

Consider a V-shaped belt, drawing the free body diagram of an element of the belt the equations of equilibrium in x and y-directions are,

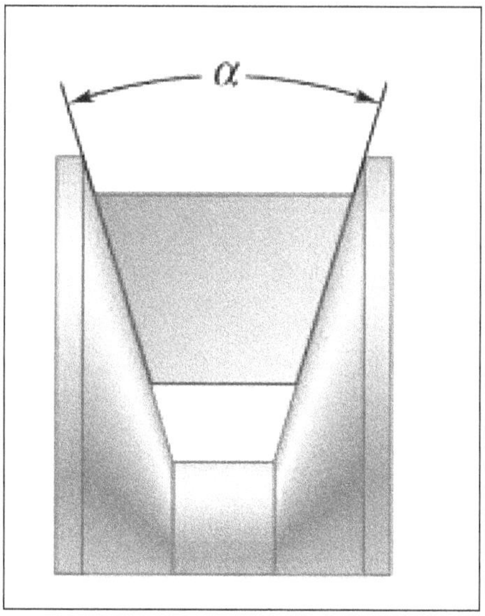

$$\sum F_y = 0 \quad 2\Delta N \ \sin \ \alpha/2 = (2T + \Delta T)\sin \ \Delta\theta/2$$

$$\sum F_X = 0 \quad 2\mu_s \Delta N = \Delta T \cos \ \Delta\theta/2$$

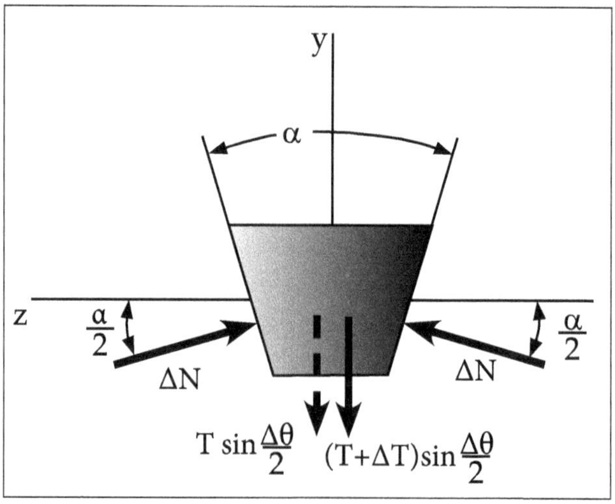

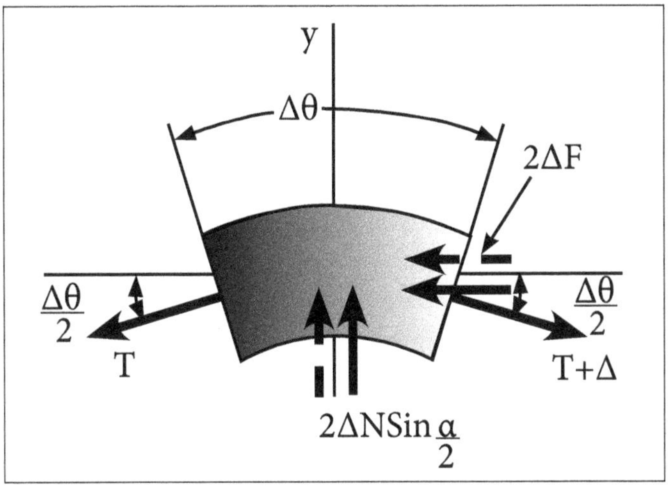

$$\Delta T\cos\Delta\theta/2 = \mu_s \frac{(2T+\Delta T)\sin\Delta\theta/2}{\sin\alpha/2}$$

$$\frac{\Delta T}{\Delta\theta}\cos\Delta\theta/2 = \frac{\mu_s}{\sin\alpha/2}\frac{(2T+\Delta T)\sin\Delta\theta/2}{\Delta\theta/2}$$

As $\Delta\theta \to 0$, the equation can be reduced,

$$dT/d\theta = \mu_s T/\sin\ \alpha/2$$

Or,

$$dT/T = \mu_s d\theta/\sin\ \alpha/2$$

Integrating and obtained,

$$T_2/T_1 = e^{\mu s\beta/\sin\ \alpha/2}$$

For $\alpha - \Pi$, $\sin\ \alpha/2 = 1$, it will reduce to the flat belt equation.

Limiting Friction

When two bodies that are not moving relative to each other are in contact, the friction force acting between their surfaces is known as static friction or limiting friction. This friction force will prevent one of the bodies from moving over the other or sliding down a slopped surface, except an applied force is greater than it.

Static friction F_s, also known as limiting friction, is the maximum friction force that is produced and which must be overcome for a body to move over or slide down the surface of another body.

Laboratory Experiment to Demonstrate Static Friction

Experiments can be carried out in the lab to show the presence of static friction on the surfaces of two unmoving bodies as follows:

Consider a body A on a plane surface and a force is gradually and increasingly applied on it using a spring balance, as shown in the figure below:

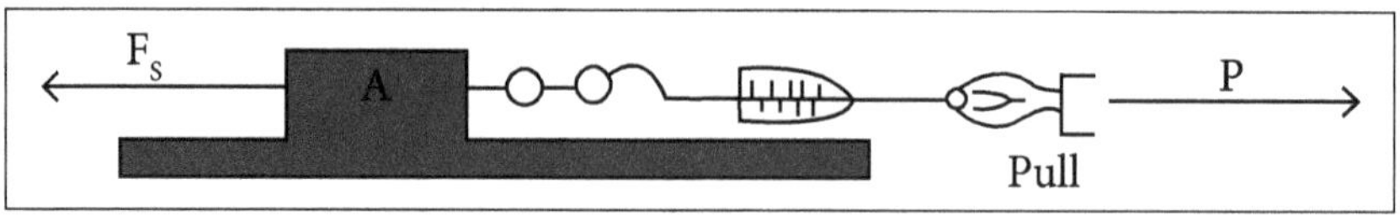

At any instant, the friction force F between the surfaces adjusts itself to be equal and opposite to P, so that the body is in equilibrium.

As the force P is gradually increased, a point is reached when the body A is just about to move. The body is said to be in limiting equilibrium at this stage and the value of P, which is noted by the spring balance represents the value of the maximum frictional force, which is also known as the static or limiting friction, F_s that is acting on the surfaces.

If P is increased further, beyond the static or limiting friction, the body begins to move with steady speed. The frictional force now acting is known as Kinetic Friction, also known as Sliding or Dynamic Friction, Fd. Dynamic friction is usually found to be less than static friction.

A good example of the action of static friction is at contact between a car's tire and the ground. Even when the car is in motion, the section of the tire in contact with the ground is not actually moving relative to the ground, therefore static friction is present and enables the car's wheel to roll on the ground without losing balance.

Angle of Friction

Consider that a body A of weight (W) is resting on a horizontal plane B, as shown in figure:

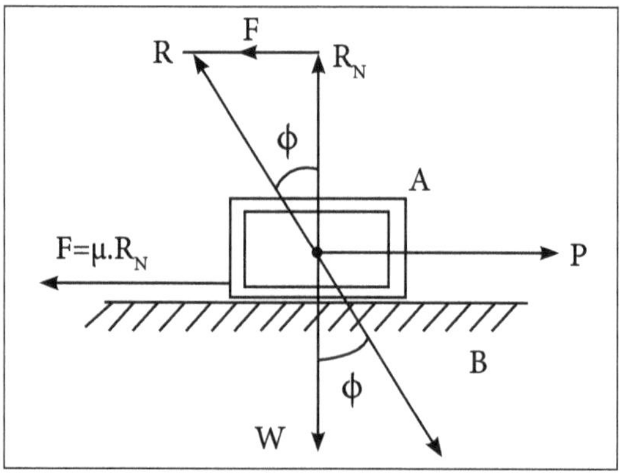

Limiting angle of friction.

If a horizontal force P is applied to the body, no relative motion can take place until the applied force P is equal to the force of friction F, acting opposite to the direction of motion. The magnitude of this force of friction is $F = \mu.W = \mu.R_N$.

Where R_N is the normal reaction.

In the limiting case, when the motion just begins, the body can be in equilibrium under the action of the following three forces:

- Weight of the body (W).
- Applied horizontal force (P).
- Reaction (R) between the body A and the plane B.

Problems

1. A 10 m long ladder rests on a horizontal floor and leans against a vertical wall. If the co-efficients of friction between the ladder and the floor and between the ladder and the wall are respectively $\mu_f = 0.3$ and $\mu_W = 0.15$, let us determine the angle of inclination of the ladder with the floor at the point of impending motion.

Solution:

The free-body diagram of the ladder is shown below.

Given:

Length of ladder = 10 m

$\mu_f = 0.3$

$\mu_W = 0.15$

To find:

Angle of inclination,

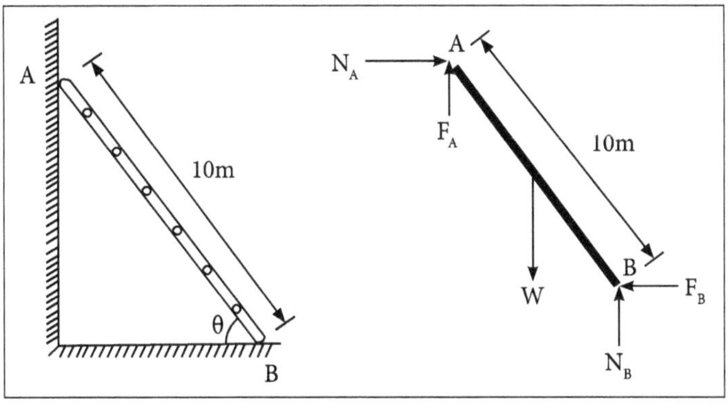

At the point of impending motion, applying the equations of equilibrium along the X and Y directions,

$$\sum F_x = 0,\ N_A - F_B = 0$$

Since,

$$F_B = \mu_f N_B$$

$$N_A - 0.3\ N_B = 0 \qquad ...(a)$$

$$\sum F_Y = 0$$

$$F_A + N_B - W = 0$$

Since,

$$F_A = \mu_W N_A,$$

$$\mu_W N_{A+}\ N_B - W = 0$$

$$0.15 N_{A+}\ N_B - W = 0 \qquad ...(b)$$

Substituting the value of N_B from equation (a) in the above equation, we get

$$W = 0.15\ N_A + (N_A/0.3)$$

$$W = 3.48\ N_A \quad ...(c)$$

We can take the moment about either A or B as both have only one unknown.

Taking the moment about B and applying the condition of equilibrium,

$$\sum M_B = 0,$$

$$-[F_A \times 10\ \cos\theta] - [\ N_A \times 10 \sin\theta] + [W \times 5 \cos\theta] = 0$$

$$-[\mu_W N_A \times 10 \cos\theta] - [\ N_A \times 10 \sin\theta] + [3.48 N_A \times 5 \cos\theta] = 0$$

$$-1.5\ \cos\ \theta - 10\ \sin\theta + 17.4\ \cos\theta = 0$$

$$15.9\ \cos\theta = 10\ \sin\ \theta$$

$$\tan\ \theta = 1.59$$

$$\theta = 57.83°$$

The below figure shows a ladder AC resting on the ground and leaning against a wall.

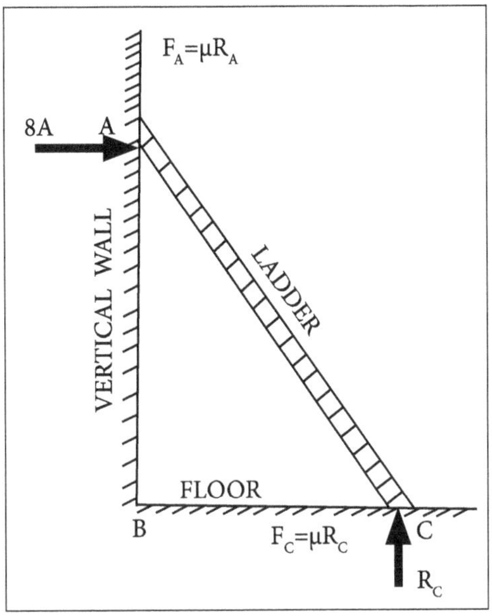

Let,

R_A = Reaction at A

R_c = Reaction at C

F_A = Force at friction at A

$= \mu R_A$

F_C= Force of friction at C

$= \mu R_c$.

Due to the self-weight of the ladder or when some man stands on the ladder, the upper end A of the ladder tends to slip downwards and hence the force of friction between the ladder and the vertical wall $F_A = \mu R_A$ will be acting upwards as shown in figure. Similarly, the lower end C of the ladder will tend to move towards right side and hence a force of friction between ladder and floor $F_c = \mu R_c$ will be acting towards left.

For the equilibrium of the system, the algebraic sum of the horizontal and vertical components of the forces must be equal to zero. Also the moments of all the forces about any point must be equal to zero.

Note: If the vertical wall is smooth, then there will be no force of friction between the ladder and the vertical wall.

2. A ladder 5 m long weighing 200 N is resting against a wall at an angle of 60° to the

horizontal ground. A man weighing 500 N climbs the ladder. At what position along the ladder from bottom does he induce slipping. The co-efficient of friction for both the wall and the ground with the ladder is 0.2.

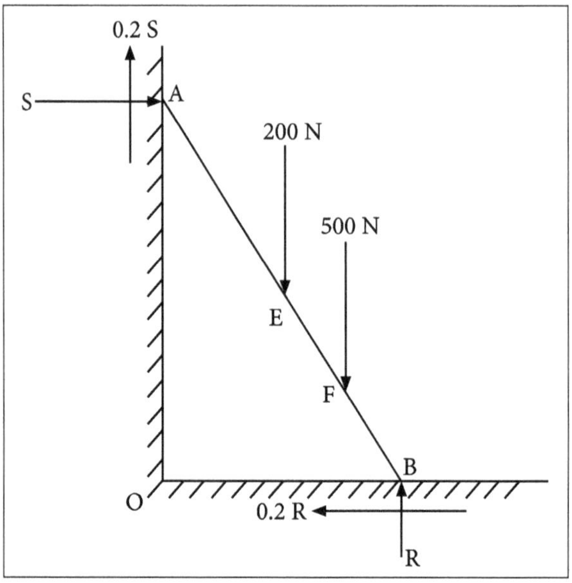

Solution:

Given:

Length of ladder = 5 m

Weight of ladder = 200 N

Angle = 60°

Weight of man = 500 N

Co-efficient of friction = 0.2

To find:

The position along the ladder from bottom,

Let the ladder be at the point of sliding when the man is at a distance x metres from the foot of the ladder.

Let F be the position of the man.

BF = x, BE = AE = 2.5m

Let the normal reactions at the floor and the wall be R and S. Friction at the floor and the wall will be 0.2R and 0.2S respectively.

Resolving the forces on the ladder horizontally and vertically,

$$S = 0.2R \quad ...(1)$$

$$R + 0.2S = 700N \quad ...(2)$$

From equations (1) and (2), we get,

$$R = 673.08 \text{ N}$$

$$S = 134.62 \text{ N}$$

Taking moments about the lower end of the ladder,

$$200 \times 2.5 \cos 60° + 500 \times x \cos 60° = S \times 5 \sin 60° + 0.2 \text{ S} \times 5 \cos 60°$$

$$250 + 250x = 2.5 \text{ S} + 0.5 \text{ S}$$

$$250 + 250x = \text{S} (2.5 + 0.5)$$

$$250 + 250x = 134.62 (2.5 + 0.5)$$

$$250 + 250x = 650.23$$

$$x = 1.60 \text{ m}$$

3. A Two blocks A and B of mass 50 kg and 100 kg respectively are connected by a string C which passes through a frictionless pulley connected with the fixed wall by another string D as shown in figure. Let us determine the force P required to pull the block B and also calculate the tension in the string D. Take co-efficient of friction at all contact surfaces as 0.3.

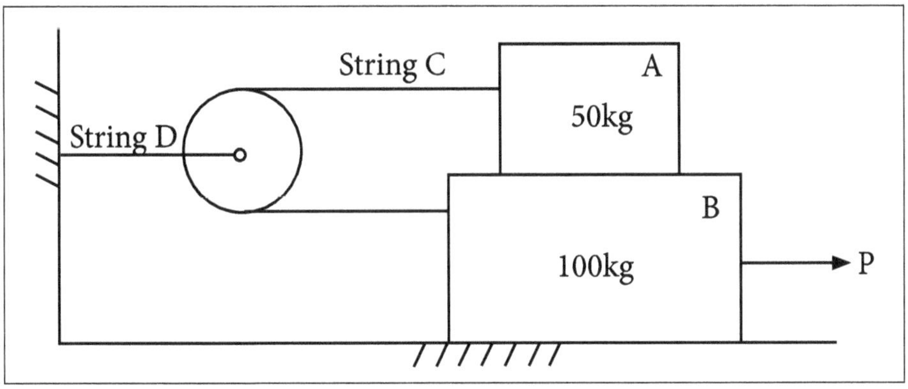

Solution:

Given:

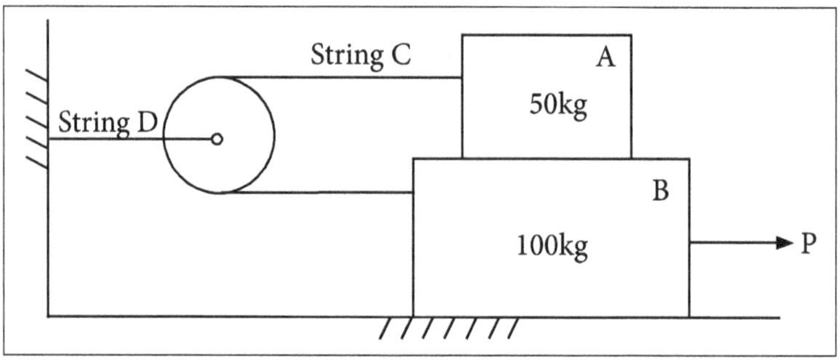

$m_A = 50kg$

$m_B = 100kg$

Co-efficient of friction = 0.3

To find:

Force:

Formula to be used:

$F_2 = v_2 \times N_2$

$P = T_1 + F_1 + F_2$

Tension in string $D' = T_2$

Free Body Diagram at Block A and B

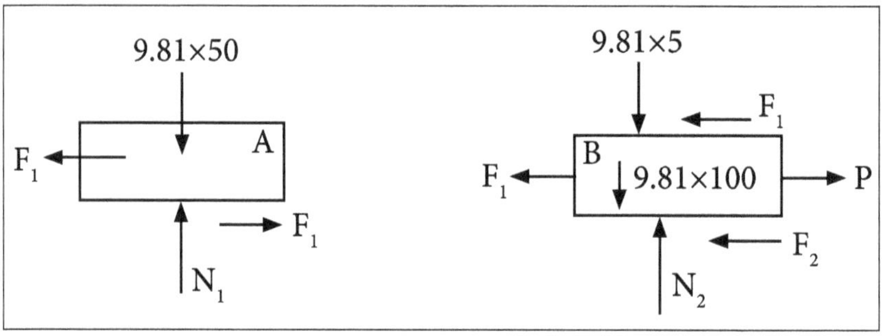

Consider Block A

$N_1 = 50 \times 9.81 = 490 \text{ N}$

Where, F_i - Friction force on block A.

Pulling force acting on the block A is F_1.

$$T_1 = F_1 = 0.3 \times 490.5 = 147.15N.$$

Consider Block B

$$\left(\sum V = 0\right)$$

$$N_2 = (50 + 100) \times 9.81 = 1471.5$$

$$N_2 = 1471.5\ N$$

We Know that,

$$F_2 = v_2 \times N_2$$

$$P = T_1 + F_1 + F_2$$

$$P = 147.15 + 147.15 + (0.3 \times 1471.5)$$

$$P = 735.75\ N$$

Tension in string D′ $= T_2$

$$T_2 = T_1 + T_1$$

$$= 147 + 147 = 294.3$$

Result:

Force in the string D = 294 N.

5. The static co-efficient of friction μ_s between the block shown in figure having a mass of 75 kg and the surface. Let us calculate the magnitude and direction of the friction force if the force P applied is inclined at 45° to the horizontal and $\mu_s = 0.30$.

Solution:

Given:

Mass = 75 kg

P is inclined at 45°

$\mu_s = 0.30$

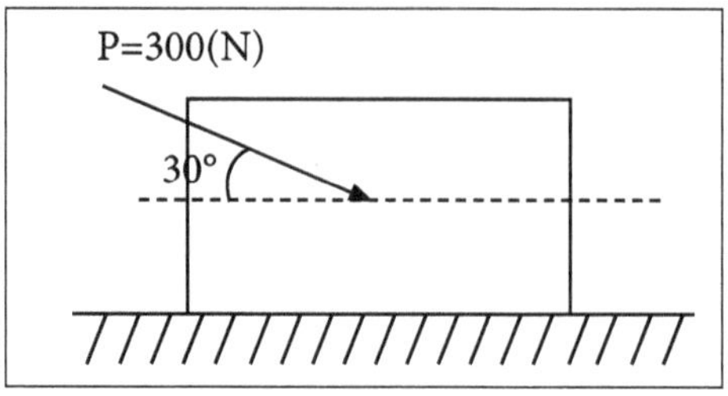

To find:

Case 1:

Mass of the block = 75kg.

Weight of the block = 75 x 9.81 = 735.75 N

Force applied, P = 300 N.

Inclination of the force with horizontal = 30°.

Co-efficient of friction, μ_s = ?

Case 2:

Inclination of the force = 45°.

Co-efficient to friction, μ_s = 0.30.

Magnitude of frictional force = ?

Direction of frictional force = ?

Formula to be used:

$F = \mu N_R$

Case 1:

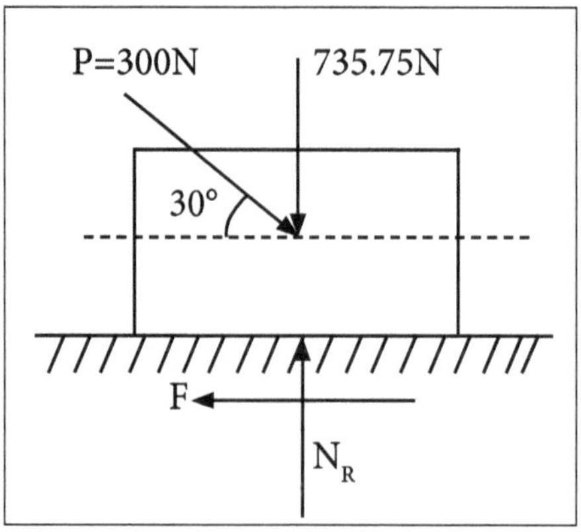

$$N_R - 300 \sin 30° - 735.75 = 0$$

$$N_R = 885.75 \text{ N}$$

$$F = \mu N_R = 885.75 \mu$$

$$F = 885.75 \mu$$

$$300 \cos 30° - F = 0$$

$$300 \cos 30° - 885.75\mu = 0$$

$$\mu = 0.293$$

Case 2:

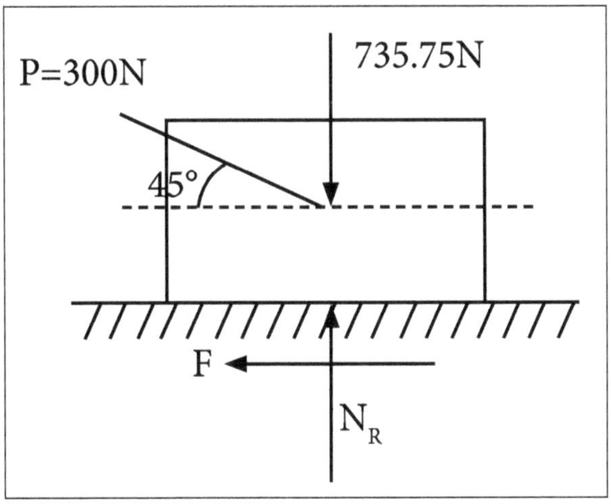

$$N_R - 300 \sin 45° - 735.75 = 0$$

$$N_R = 947.88 \text{ N}$$

$$F = \mu N_R = 0.3(947.88)$$

$$F = 284.36 \text{ N}$$

Result:

$$\mu = 0.293$$

$$F = 284.36\text{N}$$

2.3.2 Angle of Repose

Angle of repose is defined as the minimum angle made by an inclined plane with horizontal such that an object placed on the inclined surface just begins to slide.

Relation between Angle of Friction and Angle of Repose

Let us consider a body of mass 'm' resting on a plane.

Also, consider when the plane makes 'θ' angle with the horizontal, the body just begins to move.

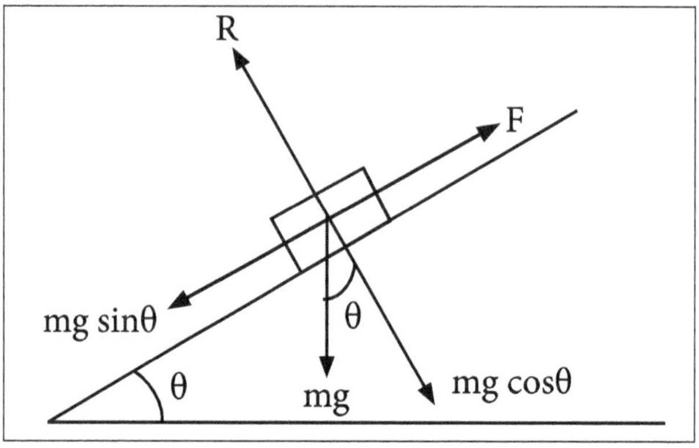

Let 'R' be the normal reaction of the body and 'F' be the frictional force.

$$mg\sin\theta = -F \quad ...(i)$$

$$mg\cos\theta = -R \quad ...(ii)$$

Dividing equation (i) by (ii),

$$\frac{mg\sin\theta}{mg\cos\theta} = \frac{-F}{-R}$$

Or, $\text{Tan}\theta = \frac{F}{R}$

Or, Tanθ = μ, where 'μ' is the co-efficient of friction

Or, Tanθ = Tanα (tanα = μ)

where 'α' is the angle of friction

$$\theta = \alpha$$

Angle of repose is equal to angle of friction.

2.3.3 Impending Motion on Horizontal and Inclined Planes

Impending motion is known to exist because the system is on the verge of slipping. The equilibrium equations and basic equation is $F_{max} = \mu_s$.

Impending Motion on Horizontal Plane

Consider a rigid body of weight W placed on a surface for which the coefficient of friction is p and angle of friction is S. A force P applied horizontally impends the motion of the body. By drawing the FBD as shown in figure (b) and applying the equations of equilibrium, the magnitude of force P can be obtained as follows:

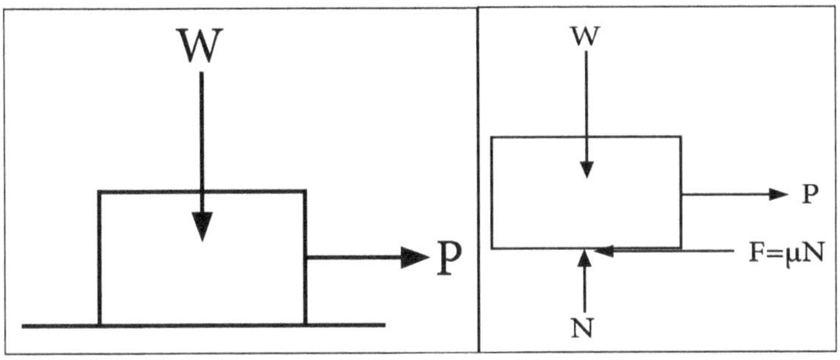

(a) Body on horizontal plane under horizontal pull P (b) Free body diagram.

$$\sum F_X = 0 \rightarrow (+), \leftarrow (-) P - \mu N = 0 \qquad ...(1)$$

$$\sum F_y = 0 \uparrow (+), \downarrow (-) N - W = 0 \qquad ...(2)$$

Simplifying equations (1 and 2),

$$P = \mu W = W \tan \phi$$

$$\mu = \tan \phi$$

Suppose the force P is inclined to horizontal at an angle θ as shown in figure in that case the value of P can be determined as follows:

Applying conditions of equilibrium,

$$\sum F_X = 0 \rightarrow (+), \leftarrow (-) P \cos \theta - \mu N = 0 \qquad ...(3)$$

$$\sum F_y = 0 \uparrow (+), \downarrow (-) N + P \sin \theta - W = 0 \qquad ...(4)$$

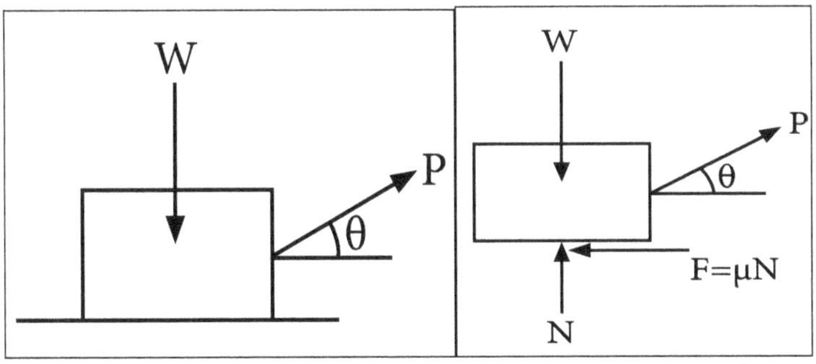

(a) Body on horizontal plane under inclined pull P (b) Free body diagram.

Substituting the value of N from 4 into 3 yields,

$$P\cos\theta = \mu(W - P\sin\theta)$$

$$P(\cos\theta + \mu\sin\theta) = \mu W$$

Hence,

$$P = \frac{\mu W}{\cos\theta + \mu\sin\theta}$$

If μ = tan Φ is substituted then,

$$P = \frac{W\tan\phi}{\cos\theta + \tan\phi\sin\theta}$$

$$P = \frac{W\sin\phi}{\sin(\theta + \phi)}$$

Impending Motion on Inclined Plane

Case 1: A body of weight W placed on an inclined plane at an angle a to the horizontal is subjected to a horizontal force P [Figure (a)]. Coefficient of friction between the body and plane is p. The force P required to cause the impending motion of the body up the inclined plane can be determined by drawing the free body diagram (FBD) as shown in figure (b).

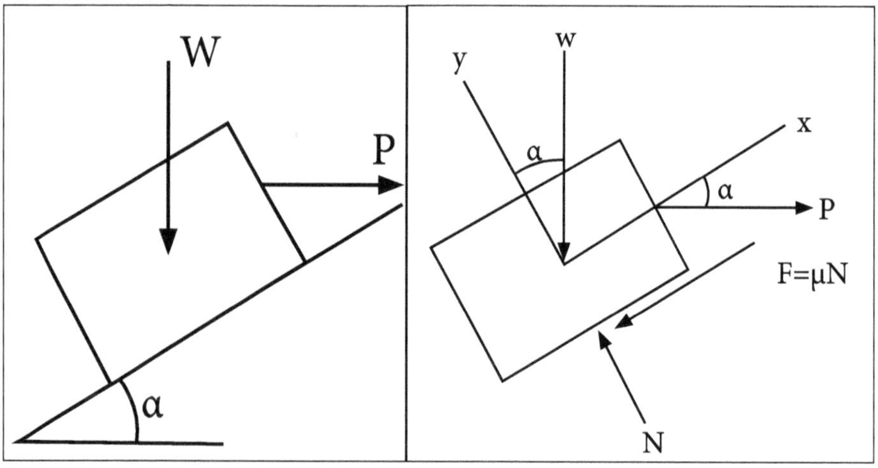

(a) Body on inclined plane under horizontal pull P (b) Free body diagram.

Considering the rectangular coordinates along the plane and normal to the plane and applying equations of equilibrium.

$$\sum F_x = 0 \;\overset{(+)}{\underset{(-)}{\rightleftarrows}}\; -\mu N + P\cos\alpha - W\sin\alpha \qquad ...(5)$$

$$\sum F_y = 0 \;\overset{(-)}{\underset{(+)}{\rightleftarrows}}\; -N - P\sin\alpha W\cos\alpha = 0 \qquad ...(6)$$

Simplifying,

$$-\mu(P\sin\alpha + W\cos\alpha) + P\cos\alpha - W\sin\alpha = 0$$

$$P(\cos\alpha - \mu\sin\alpha) = W(\sin\alpha + \mu\cos\alpha)$$

Therefore,

$$P = W\left[\frac{\sin\alpha + \mu\cos\alpha}{\cos\alpha - \mu\sin\alpha}\right] = W\left[\frac{\tan\alpha + \mu}{1 - \mu\tan\alpha}\right]$$

If $\mu = \tan\phi$, when φ = Angle of friction then $P = W\tan(\alpha + \varphi)$.

On the other hand, the force P required to prevent the block from sliding down can be determined by simply changing the direction of frictional force upward instead of downwards.

Because in this case, the body will be on the verge of motion down the plane. The force P can be obtained similar to the way as above except the term (μN) in equation (5) will have +ve sign. The magnitude of P will be,

$$P = W\left[\frac{\sin\alpha - \mu\cos\alpha}{\cos\alpha + \mu\sin\alpha}\right] = W\left[\frac{\tan\alpha - \mu}{1 + \mu\tan\alpha}\right]$$

or,

$$P = W\tan(\alpha - \Phi)$$

Case 2: A body of weight W placed on an inclined plane at an angle α to the horizontal is subjected to a force P that is shown in figure (a), parallel to the plane and up the plane. p is the coefficient of friction between the body and plane.

To determine the magnitude of P required impending the body up the plane. The FBD is drawn as shown in figure (b) and the equations of equilibrium are applied considering the rectangular directions as shown.

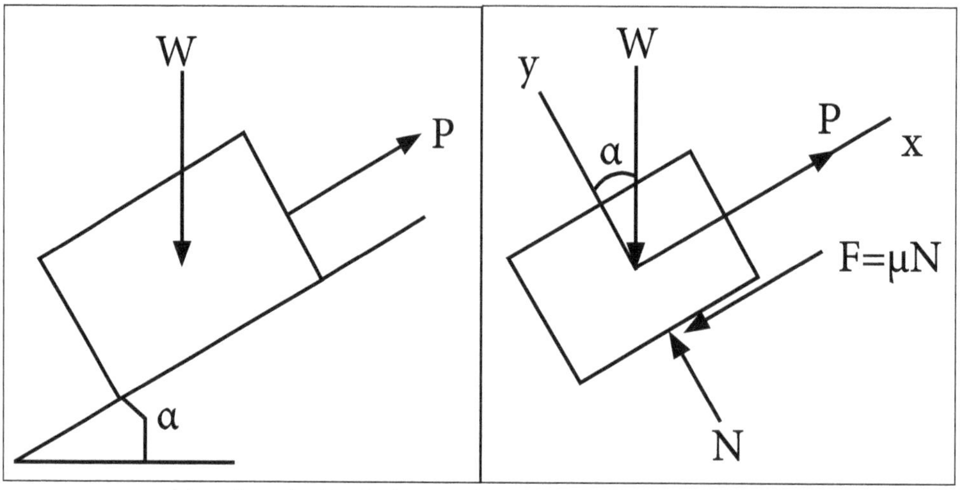

(a) Body on inclined plane under pull (b) Free body diagram.

$$F = \mu Y$$

$$\Sigma F_x = 0 \underset{(-)}{\overset{(+)}{\rightleftarrows}} -\mu N + P - W \qquad ...(7)$$

$$\Sigma F_y = 0 \underset{(+)}{\overset{(-)}{\rightleftarrows}} N - W\cos\alpha = 0 \qquad ...(8)$$

Hence,

$$N = W\cos\alpha \text{ and} - \mu(w\cos\alpha) + P - W\sin\alpha = 0.$$

Therefore,

$$P = W(\sin\alpha + \mu\cos\alpha) \qquad ...(9)$$

Substituting p = tan φ Equation (9) reduces to,

$$P = W(\sin\ \alpha + \tan\phi\cos\alpha)$$

$$P = W\left(\frac{\sin(\alpha - \phi)}{\cos\phi}\right) \qquad ...(10)$$

Further, if it is required to find the force P to prevent the block from sliding downwards, the direction of frictional force in FBD is reversed as the motion becomes impending downwards. The corresponding term in equation (7) is taken with +ve sign. Hence, equation (8) becomes,

$$P = W(\sin\ \alpha - \mu\ \cos\alpha) \qquad ...(11)$$

Similarly, equation (10) reduces to,

$$P = W\left[\sin\left(\frac{\alpha-\phi}{\cos\phi}\right)\right] \quad ...(12) \qquad ...(12)$$

Case 3: A body of weight W placed on an inclined plane at an angle α to the horizontal having coefficient of friction p. A force P is applied to the body inclined at an angle θ to the plane.

Considering the FBD for the case when P causes impending motion up the plane and applying the equations of equilibrium yields,

$$\sum F_x = 0 \xrightleftharpoons[(-)]{(+)} -\mu N + P\cos\theta - W\sin\alpha = 0$$

$$\sum F_y = 0 \xrightleftharpoons[(+)]{(-)} N + P\sin\theta - W\cos\alpha = 0$$

$$N = W\ \cos\ \alpha - P\sin\theta \qquad ...(14)$$

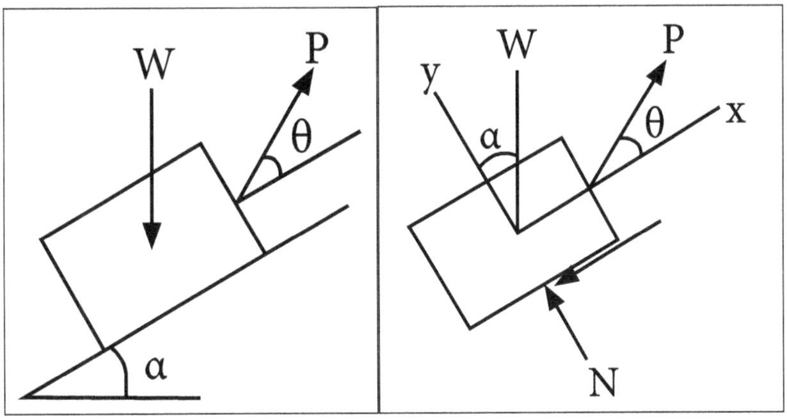

(a) Body on inclined plane under pull P (b) Free body diagram.

Substituting in equation (14),

$$-\mu(W\ \cos\ \alpha - P\ \sin\ \theta) + P\ \cos\ \theta - W\ \sin\ \alpha = 0$$

$$P(\cos\ \theta + \mu\ \sin\ \theta) = W(\sin\ \alpha + \mu\ \cos\ \alpha)$$

Therefore,

$$P \quad W\left[\frac{\sin\alpha - \mu\cos\alpha}{\cos\theta + \mu\sin\theta}\right] \qquad ...(15)$$

Putting p = tan φ, equation (15) yields,

$$P = W\left[\frac{\sin(\alpha-\phi)}{\cos(\theta-\phi)}\right] \quad ...(16)$$

When the force P applied is just sufficient to prevent the block from sliding downwards it can be determined by reversing the direction of force of friction. In that case the force P becomes,

$$P = W\left[\frac{\sin\alpha-\mu\cos\alpha}{\cos\alpha+\mu\sin\alpha}\right] \quad ...(17)$$

Or,

$$P = W\left[\frac{\sin(\alpha-\phi)}{\cos(\theta-\phi)}\right] \quad ...(18)$$

2.3.4 Numerical Problems on Single and Two Blocks on Inclined Planes

1. Let us calculate the minimum force needed to dislodge a block of mass m resting on an inclined plane of slope angle α, if the coefficient of friction is μ. Also let us investigate the cases when,

Solution:

a) $\alpha = 0$;

b) $0 < \alpha <$ arc tan μ.

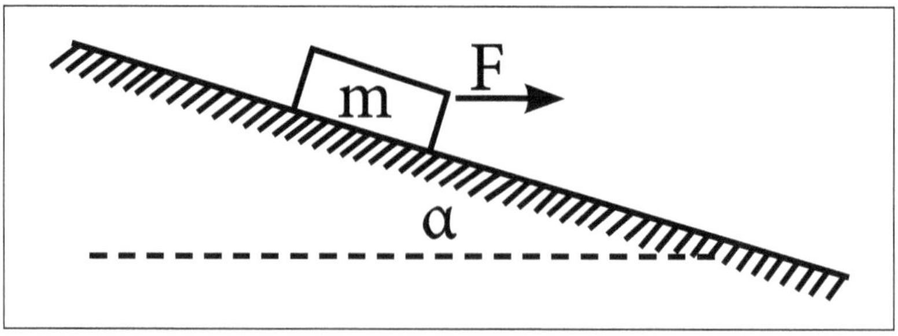

Force balance can sometimes be resolved vectorially without projecting anything onto the axes or rather its following generalization, turns out to be of use:

If a body is on the verge of slipping, then the sum of the friction force and the reaction force is angled by an arc tan μ from the surface normal.

2. A block rests on an inclined surface with slope angle α. The surface moves with a horizontal acceleration which lies in the same vertical plane as a normal vector to the surface. Let us determine the values of the coefficient of friction μ that allow the block to remain still.

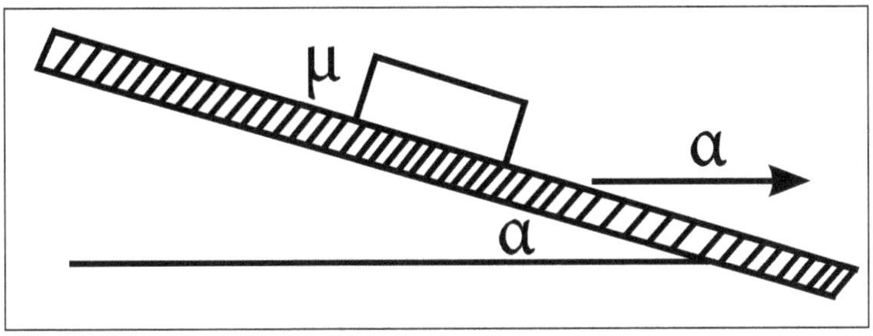

Solution:

Many problems become very easy in a non-inertial translationally moving reference frame.

To Clarify

In a translationally moving reference frame, we can re-establish Newton's laws by imagining that every body with mass m is additionally acted on by an inertial force m.a where a is the acceleration of the frame of reference.

Note that the fictitious force is totally analogous to the gravitational force and their equivalence is the corner-stone of the theory of general relativity.

The net of the inertial and gravitational forces is usable as an electro gravitational force.

3

Analysis of Non-concurrent Force Systems

3.1 Resultants and Equilibrium Composition of Coplanar and Non-concurrent Force System

Resultant of Non-concurrent Force Systems

A system of non-concurrent forces can be reduced to:

- A single moment if resultant force is zero.
- A force-couple system.
- A single resultant force with a specified line of action.

The line of action can be specified by the perpendicular distance of resultant force from a given point or x/y intercept on co-ordinate axes or intercept on any inclined axis.

The parallel force systems are also non-concurrent force systems and hence can be reduced in the same way.

The resultant force can be obtained using resolution of forces into rectangular components. The line of action of resultant force can be obtained using Varignon's theorem.

Whenever a number of forces are acting on a body, it is possible to find a single force, which can produce the same effect as that produced by the given forces acting all together. Such a single force is called as resultant force or resultant.

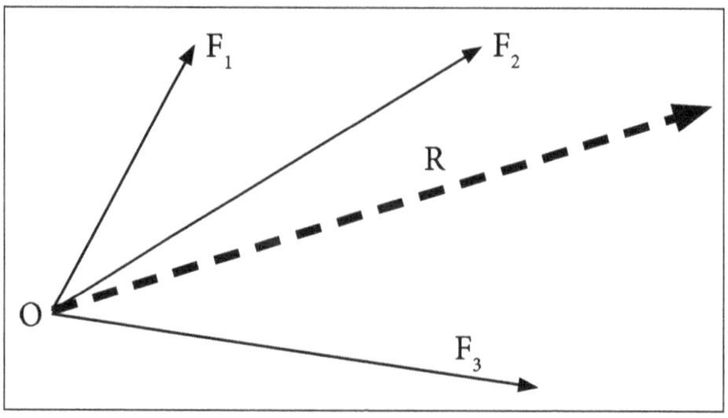

In the above figure R can be said to be the resultant of the given forces F_1, F_2 and F_3.

The process of determining the resultant force of a given system of force is called as composition of forces. The resultant force of a given force system can be determined by analytical and graphical methods.

In analytical methods two different principles namely, method of resolution of forces and Parallelogram law of forces are adopted.

Composition of System of Forces

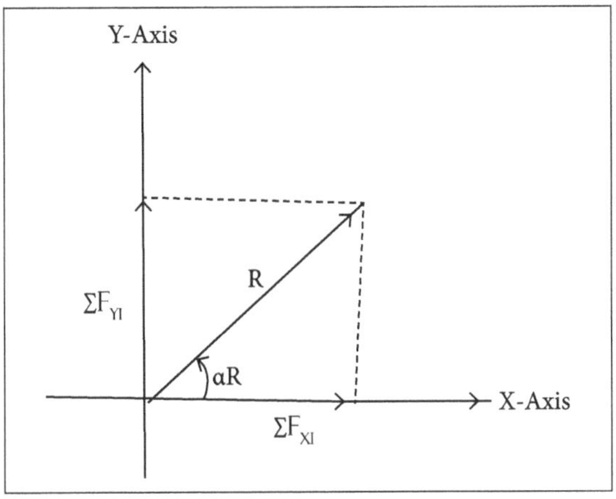

$$R = \sqrt{\left(\sum f_{xi}\right)^2 + \left(\sum f_{yi}\right)^2}$$

$$\alpha_R = \tan^{-1}\frac{\left(\sum f_{yi}\right)}{\left(\sum f_{xi}\right)}$$

Equilibrium

Equilibrium is the status of the body when it is subjected to a system of forces. We know that for a system of forces acting on a body the resultant can be determined. By Newton's 2nd Law of Motion the body then should move in the direction of the resultant with some acceleration.

If the resultant force is equal to zero it implies that the net effect of the system of forces is zero this represents the state of equilibrium. For a system of coplanar concurrent forces for the resultant to be zero,

Hence,

$$\sum f_{xi} = 0$$
$$\sum f_{yi} = 0$$

Non-concurrent Forces

Non-concurrent forces are the forces in which two or more forces whose magnitudes are equal but act in opposite directions with a common line of action.

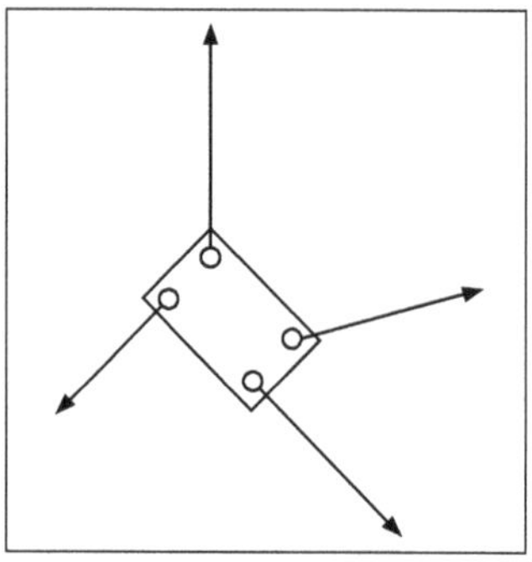

Non-Concurrent Forces.

3.1.1 Varignon's Principle of Moments

Varignon's Theorem

The principle of moments, also known as Varignon's theorem, states that the moment of any force is equal to the algebraic sum of the moments of the components of that force.

It is a very important principle that it is often used in conjunction with the Principle of transmissibility in order to solve systems of forces that are acting upon and/or within a structure.

This concept will be illustrated by calculating the moment around the bolt caused by the 100 pound force at points A, B, C, D and E in the illustration.

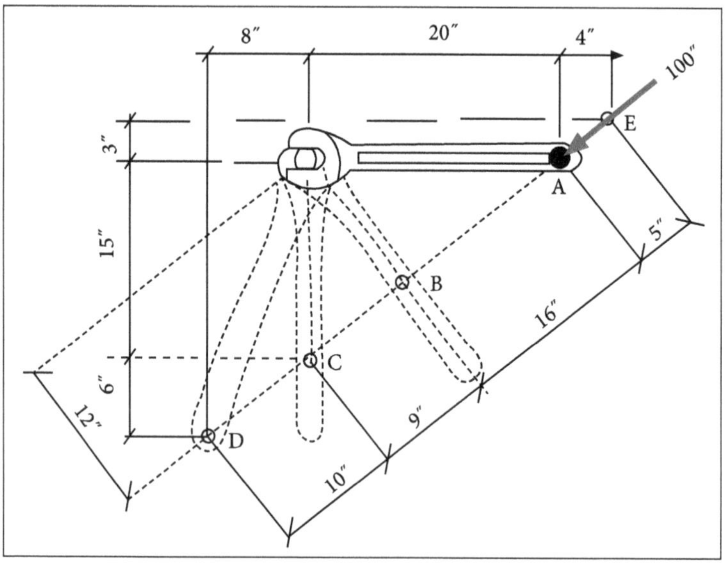

First consider the 100 pound force.

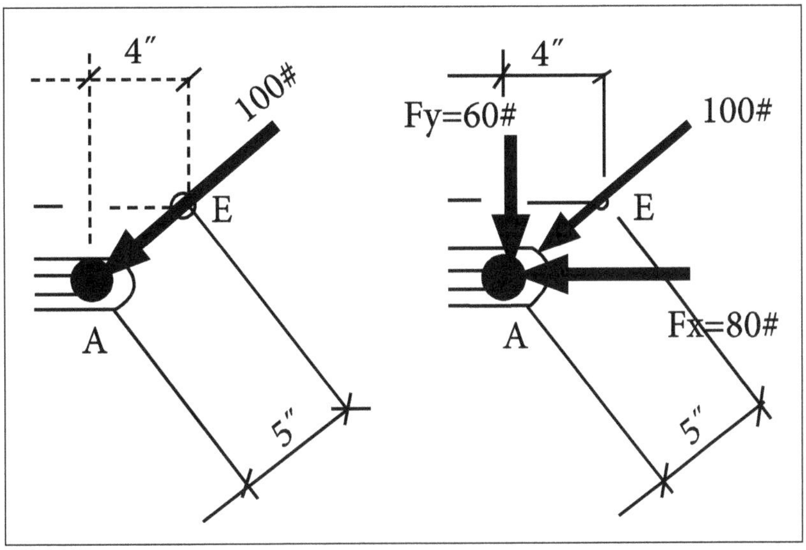

Since the line of action of the force is not perpendicular to the wrench at A, the force is broken down into its orthogonal components by inspection. The line of action of 100 pound force can be inspected to determine if there are any convenient geometries to aid in the decomposition of the 100 pound force.

The 4 inch horizontal and 5 inch diagonal measurement near point A should be recognized as belonging to a 3-4-5 triangle. Therefore, $F_x = -4/5(100 \text{ pounds})$ or -80 pounds and $F_y = -3/5(100 \text{ pounds})$ or -60 pounds.

Consider point A,

The line of action of F_x at A passes through the handle of the wrench to the bolt (which is also the center of moments).

This means that the magnitude of the moment arm is zero and therefore the moment due to F_{Ax} is zero. F_{Ay} at A has a moment arm of twenty inches and will tend to cause a positive moment.

$F_{Ay}\, d = (60 \text{ pounds})\,(20\text{inches}) = 1200$ pound-inches or 100 pound-feet

The total moment caused by the 100 pound force F at point A is 1200 pound-inches.

Consider point B,

At this point the 100 pound force is perpendicular to the wrench. Thus, the total moment due to the force can easily be found without breaking it into components.

$F_B\, d = (100 \text{ pounds})(12\text{inches}) = 1200$ pound-inches.

The total moment caused by the 100 pound force F at point B is again 1200 pound-inches.

Consider point C,

The force must once again be decomposed into components. This time the vertical component passes through the center of moments. The horizontal component F_{Cx} causes the entire moment.

F_{Cx} d = (80 pounds) (15inches) = 1200 pound-inches

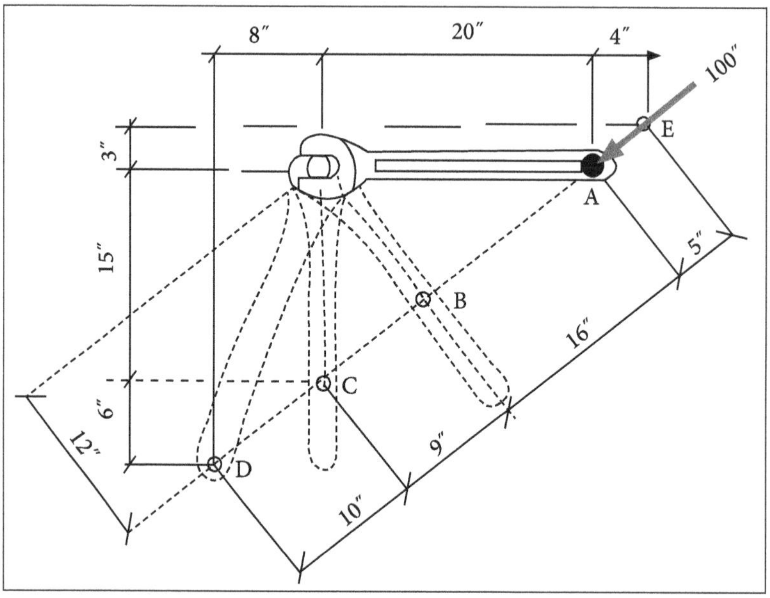

Consider point D,

The force must once again be decomposed into components. Both components will contribute to the total moment.

F_{Dx} d = (80 pounds)(21inches) = 1680 pound-inches F_{Dy} d = (60 pounds)(8inches) = −480 pound-inches

Note that the y component in this case would create a counter clockwise or negative rotation. The total moment at D due to the 100 pound force is determined by adding the two component moments. Not surprisingly, this yields 1200 pound-inches.

Consider point E,

Varignon's theorem applies even though point E is removed from the physical object. Following the same procedure as at point D we have,

F_{Ex} d = (80 pounds)(3inches) = -240 pound-inches F_{Ey} d = (60 pounds)(24inches) = 1440 pound-inches

However, this time F_x tends to cause a negative moment. Once again the total moment is 1200 pound-inches.

3.1.2 Numerical Problems on Composition of Coplanar Non-concurrent Force System

1. Let us determine the magnitude and direction of the resultant of the two forces of magnitude 12 N and 9 N acting at a point, if the angle between the two forces is 30°.

Solution:

Given:

Magnitude of two forces = 12 N and 9 N

Angle = 30°

To find:

Magnitude and direction:

$F_1 = 12\text{ N}\ F_2 = 9\text{ N}\ \alpha = 30°$

$$R = \sqrt{F_1^2 + F_1^2 + 2F_1F_2\cos\alpha}$$

$$R = \sqrt{12^2 + 9^2 + 2\times12\times9\times\cos30°}$$

$R = 20.3$ N

$$\theta = \tan^{-1}\left(\frac{F^2\sin\alpha}{F_1 + F_2\cos\alpha}\right)$$

$$\theta = \tan^{-1}\left(\frac{9\sin30°}{12 + 9\cos30°}\right)$$

$\theta = 12.81°$

2. Let us determine the magnitude of two equal forces acting at a point with an angle of 60° between them, if the resultant is equal to 30 √3 N.

Solution:

Given:

Angle = 60°

Resultant = $30\sqrt{3}$ N

To find:

Magnitude

$$F_1 = F_2 = F\text{, say}$$

$$R = 30\sqrt{3}\text{ N}, \alpha = 60°$$

$$R = \sqrt{F_1^2 + F_2^2 + 2F_1F_2\cos\alpha}$$

$$R = \sqrt{F^2 + F^2 + 2F \times F \times \cos 60°}$$

$$R = \sqrt{F^2 + F^2 + F^2}$$

$$R = \sqrt{3}F$$

$$F = 30\text{ N.}$$

3. Let us determine the resultant of the force system acting on the plate as shown in figure given below with respect to AB and AD.

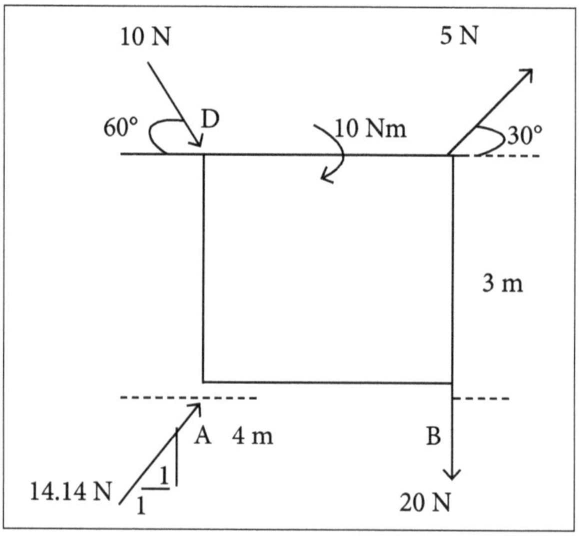

Solution:

To find:

Resultant:

$$R = \sqrt{\left(\sum F_x\right)^2 + \left(\sum F_y\right)^2}$$

$$\theta = \tan\left(\frac{\sum Fy}{\sum Fx}\right)$$

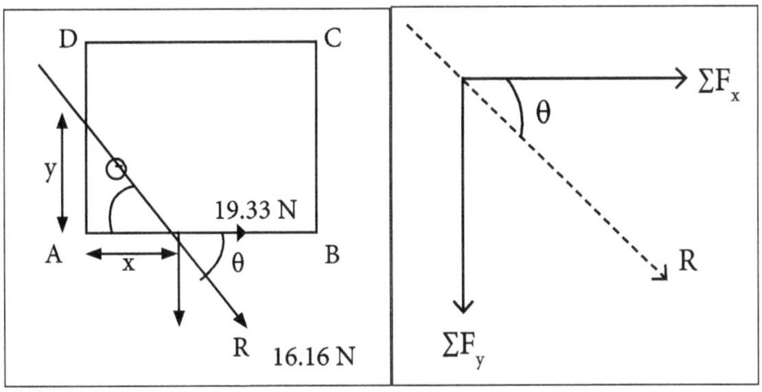

$$\sum F_x = 5\cos 30° + 10\cos 60° + 14.14\cos 45°$$

$$\sum F_x = 19.33\ \text{N}$$

$$\sum F_y = 5\sin 30° - 10\sin 60° + 14.14\sin 45°$$

$$\sum F_y = -16.16\ \text{N}$$

$$R = \sqrt{\left(\sum Fx\right)^2 + \left(\sum Fy\right)^2}$$

$$R = 25.2\text{N}$$

$$\theta = \tan^{-1}\left(\frac{\sum Fy}{\sum Fx}\right)$$

$$\theta = \tan^{-1}(16.16/19.33) = 39.89°$$

Tracing moments of forces about A and applying Varignon's principle of moments we get,

$$16.16X = (20x4) + (5\cos 30°x3) - (5\sin 30°x4) + 10 + (10\cos 60°x3)$$

$$X = 107.99/16.16 = 6.683\text{m}$$

Also,

$$\tan 39.89 = y/6.83$$

$$y = 5.586\text{m}.$$

4. The system of forces acting on a crank is shown in figure below. Determine the magnitude , direction and the point of application of the resultant force.

Solution:

Given:

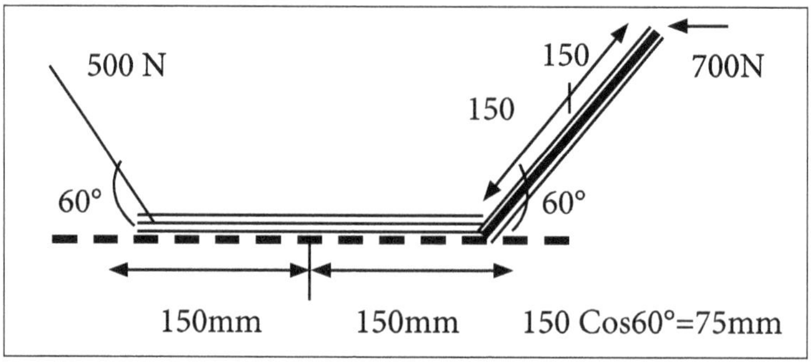

To find:

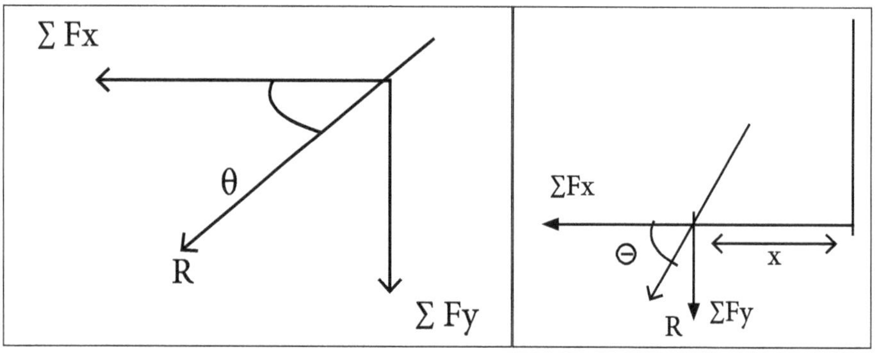

Direction, magnitude and point of application,

$$R = \sqrt{\left(\sum Fx\right)^2 + \left(\sum Fy\right)^2}$$

$$\theta = \tan^{-1}\left(\frac{\sum Fy}{\sum Fx}\right)$$

$$\sum F_x = 500\ \cos\ 60° - 700$$

$$\sum F_x = 450\ \text{N}$$

$$\sum F_y = 500\ \sin\ 60°$$

$$\sum F_y = -26.33\ \text{N}$$

$$R = \sqrt{\left(\sum Fx\right)^2 + \left(\sum Fy\right)^2}$$

$$R = 267.19 \text{ N (Magnitude)}$$

$$\theta = \tan^{-1}\left(\frac{\sum Fy}{\sum Fx}\right)$$

$$\theta = \tan^{-1}(26.33/450) = 80.30^\circ \text{ (Direction)}$$

Tracing moments of forces about θ and applying Varignon's principle of moments we get,

$-2633x = -500x \sin 60^\circ x300 - 1000 \times 150 + 1200 \times 150 \cos 60^\circ - 700 \times 300$
$\sin 60^\circ$

$X = -371769.15/-2633$

$X = 141.20$ mm from θ towards left.

3.2 Support Reaction in Beams

Support Reaction in Beams

Constraints	Type and direction of forces produced
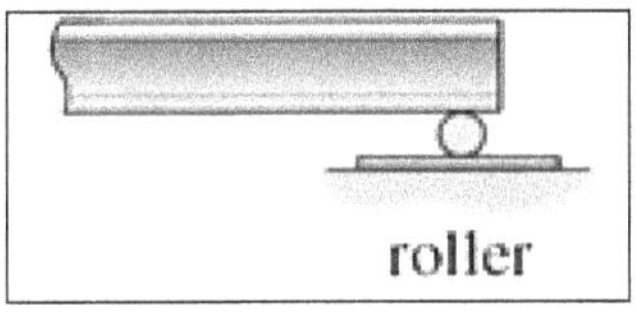 The connection point on the bar cannot move downward.	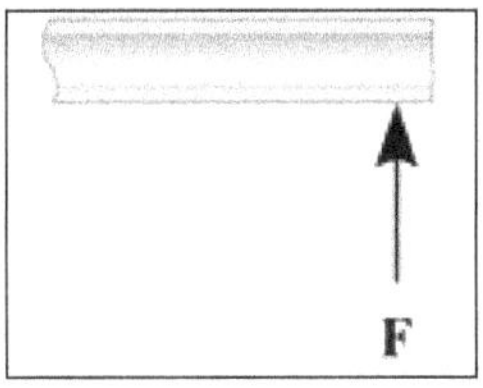
 The joint cannot move in vertical and horizontal directions.	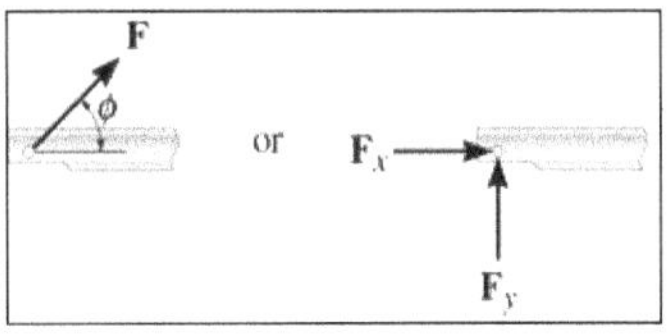
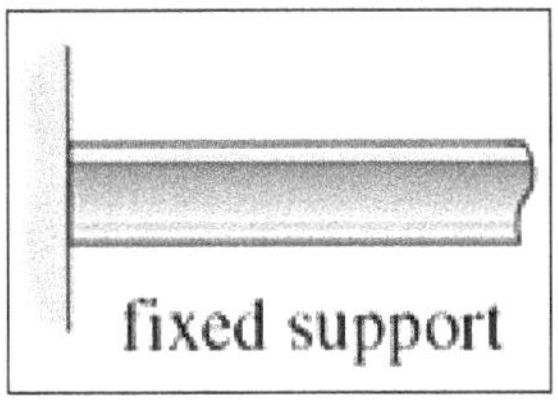 The support prevents translation in vertical and horizontal directions and also rotation, Hence a couple moments is developed on the body in that direction as well.	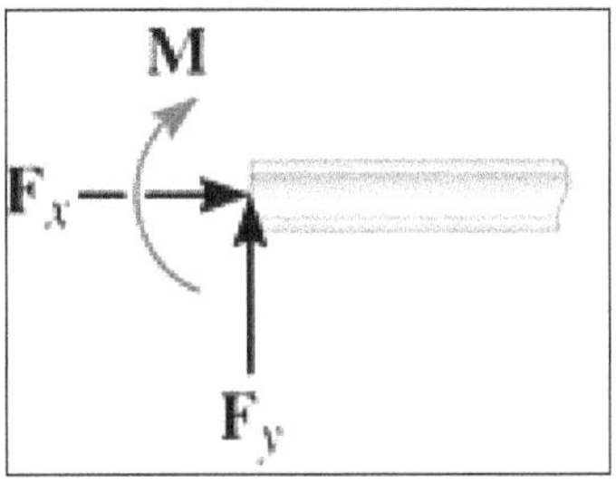

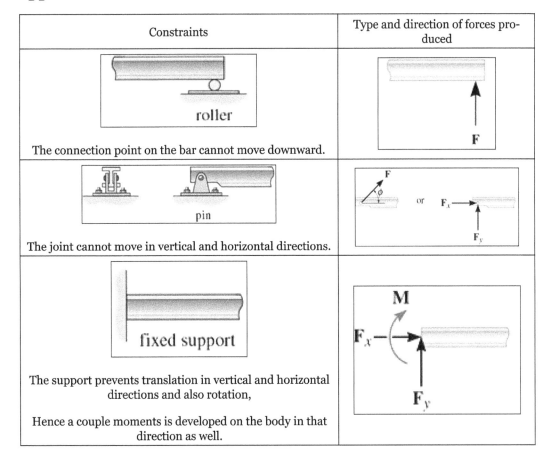

Problems

1. Let us determine the reactions of simply supported beam when a point load of 1000 kg and a uniform distributed load of 200 kg/m is acting on it. As shown in figure below.

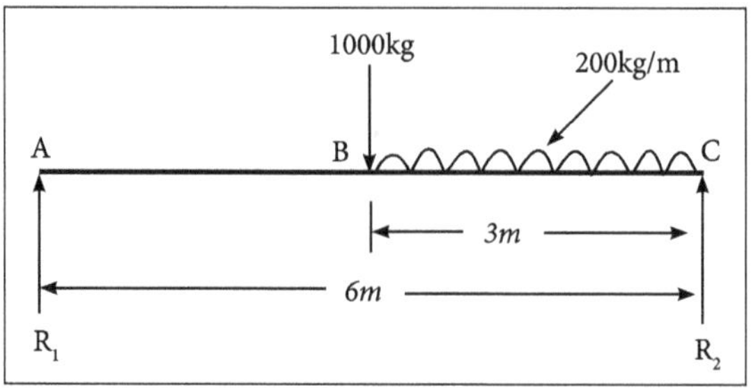

Solution:

Given:

Point load of 1000 kg

Uniform distributed load of 200 kg/m

In order to calculate reaction R_1, take moment at point C.

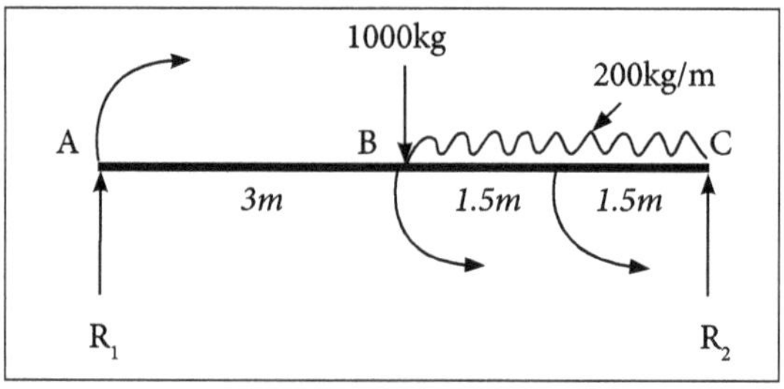

$$\sum M_c = 0$$

Clockwise moments = Anti clock wise moments

$$R_1 \text{ x } 6 = (1000\times3) + (200\times3)3/2$$

$$6R_1 = 3000 + 900 = 3900$$

$$R_1 = 3900/6 = 650 \text{ kg.}$$

$$R_1 = 650 \text{ kg}$$

For calculating R_2 i.e. reaction at point C,

$$\sum F_V = 0$$

$$R_1 + R_2 = 600 + (200 \times 4)$$

$$1300 + R_2 = 1400$$

$$R_2 = 1400 - 1300$$

$$R_2 = 100 \text{ kg}$$

3.2.1 Types of Loads and Supports

Types of Load

There are three types of load. These are:

- Point load that or concentrated load.
- Distributed load.
- Coupled load.

Point Load

Point load is that load which acts over a small distance. Because of concentration over small distance this load can be considered as acting on a point. Point load is denoted by P.

Distributed Load

Distributed load is that it acts over a considerable length or we can say "over a length which is measurable". Distributed load is measured as per unit length.

Types of Distributed Load:

Distributed load is further divided into two types,

- Uniformly Distributed load (UDL)
- Uniformly Varying load (Non-uniformly distributed load).

Uniformly Distributed Load (Udl)

Uniformly distributed load is the one whose magnitude remains uniform throughout the length.

Uniformly distributed load is usually represented by W and is pronounced as intensity of udl over the beam, slab etc.

Uniform Distributed Load to Point Load:

Conversion of uniform distributed load to point load is very simple. By simply multiplying the intensity of udl with its loading length. The answer will be the point load which can also be pronounced as Equivalent concentrated load (E.C.L). Concentric because converted load will acts at the center of span length.

Mathematically, it can be written as,

Equivalent Concentrated load = udl intensity (W) x Loading length

Uniformly Varying Load (Non - Uniformly Distributed Load)

It is that load whose magnitude varies along the loading length with a constant rate.

Uniformly varying load is further divided into two types:

- Triangular Load.
- Trapezoidal Load.

Triangular Load

Triangular load is that whose magnitude is zero at one end of span and increases constantly till the second end of the span.

Trapezoidal Load

Trapezoidal load is the one which is acting on the span length in the form of trapezoid. Trapezoid is generally form with the combination of uniformly distributed load (UDL) and triangular load.

Coupled Load

Coupled load is the one in which two equal and opposite forces acts on the same span. The lines of action of both the forces are parallel to each other but opposite in directions. This type of loading creates a couple loads.

Coupled load try to rotate the span in case one load is slightly more than the second load. If force on one end of beam acts upward then same force will acts downwards on the opposite end of beam.

Types of Supports

Types of reaction and its direction will depend upon the type of support provided:

- Friction less or smooth surface support.
- Roller support.
- Knife edge support.

- Hinged or pinned support.

Roller Support

Roller supports are free to translate and rotate along the surface upon which the roller rests. The surface may be slopped, horizontal or vertical at any angle. Roller supports are commonly located at one end of long bridges in the form of bearing pads.

This support allows the bridge structure to contract and expand with temperature changes and without this expansion the forces can fracture the supports at the banks. This support cannot provide resistance to lateral forces.

Roller support is also used in frame cranes in heavy industries as shown in figure, the support can move towards left, right and rotate by resisting vertical loads. Thus, a heavy load can be shifted from one place to another horizontally.

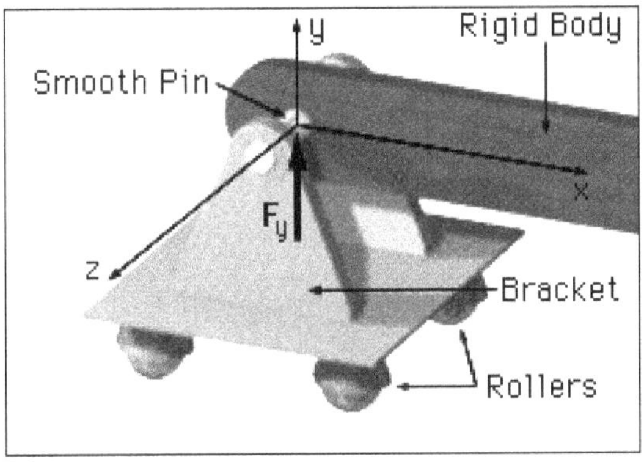

Roller support.

Hinge Supports

The hinge support is capable of resisting forces acting in any direction of the plane. This support does not provide any resistance to rotation. The vertical and the horizontal components of the reaction can be determined by using the equation of equilibrium.

Hinge support can also be used in three hinged arched bridges at the banks supports while at the center, the internal hinge is introduced. It is also used in doors to produce rotation. Hinge support reduces sensitivity to earthquake.

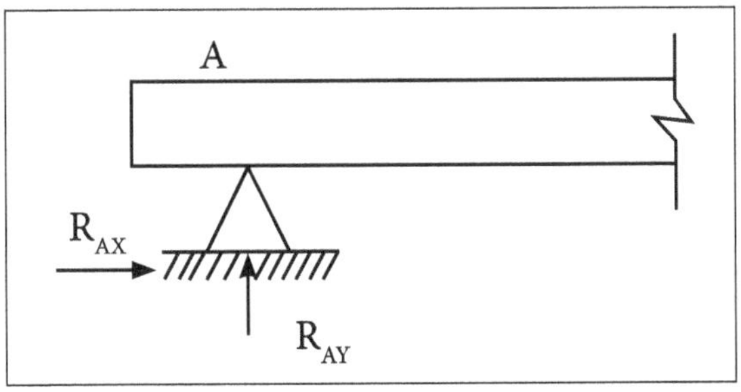

Hinge support.

Fixed Support

Fixed support can resist vertical and horizontal forces as well as moment since they restrain both translation and rotation. They are also known as rigid support, for the stability of a structure there should be one fixed support always.

A flagpole at concrete base is common example of fixed support. In RCC structures, the steel reinforcement of a beam is embedded in a column to produce a fixed support.

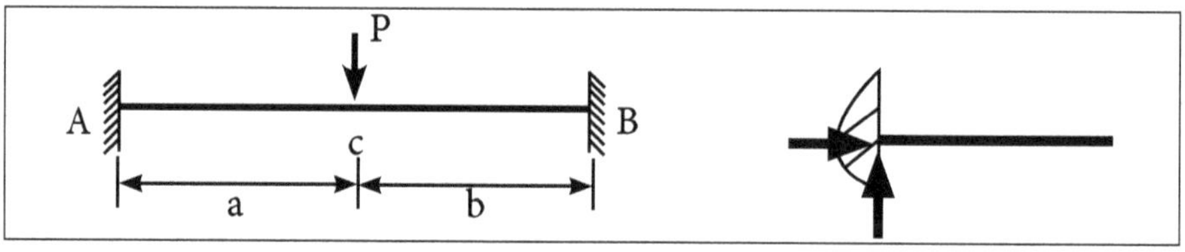

Similarly all the welded and the riveted joints in steel structure are the examples of fixed supports. Riveted connections are not very much common nowadays due to the introduction of bolted joints.

Pinned Supports

A pinned support is same as hinged support. It can resist both horizontal and vertical forces but not a moment. It allows the structural member to rotate, but not to translate in any direction. Many connections are assumed to be pinned connections even though they might resist a small amount of moment in reality.

It is very true that a pinned connection could allow rotation in one direction only; providing rotation resistance in any other direction. Human body knee is the best example of hinged support as it allows rotation in one direction only and resists lateral movements.

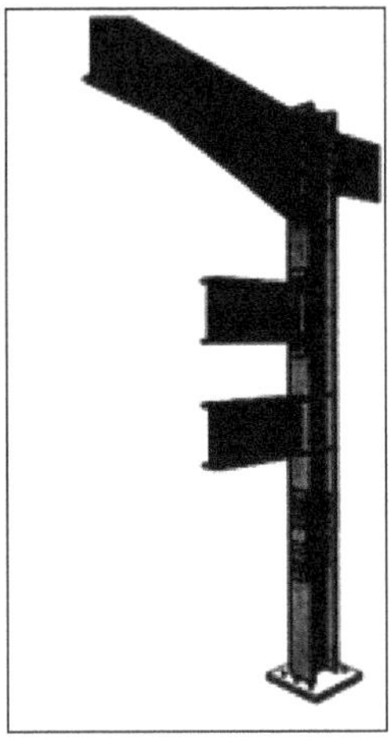

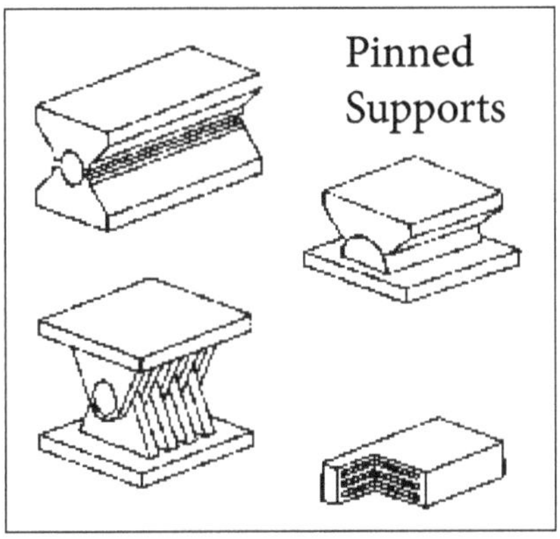

Pinned support.

Ideal pinned and fixed supports are rarely found in practice, but beams that are supported on walls or simply connected to other steel beams are regarded as pinned support. The distribution of shear force and moments is influenced by the support condition.

Internal Hinge

Interior hinges are often used for joining flexural members at points other than supports. For example in below figure two halves of an arch is joined through the internal hinge. In some cases it is intentionally introduced so that excess load breaks this weak zone rather than damaging other structural elements as shown in below image:

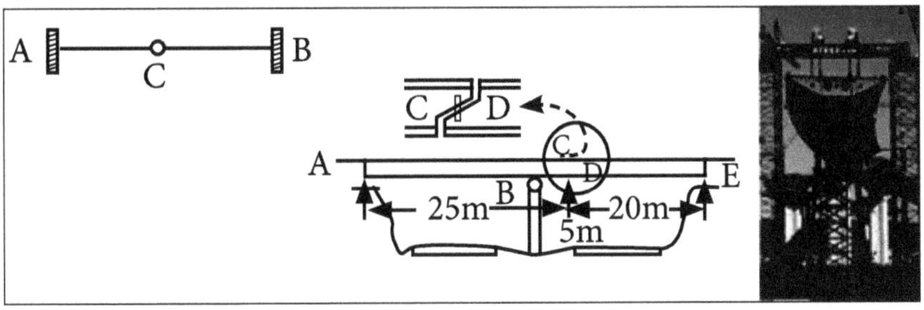

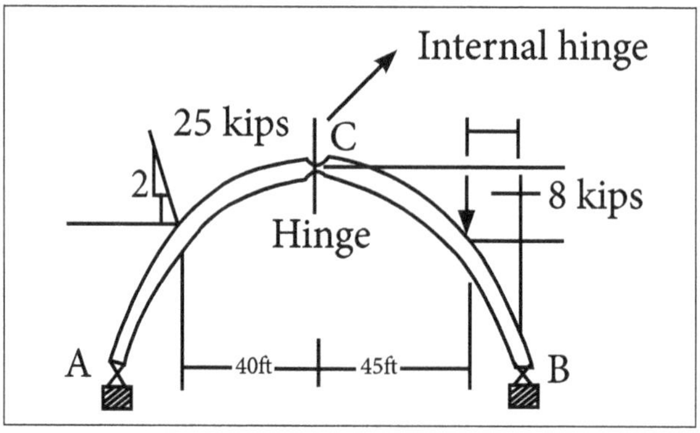

Internal hinge.

3.3 Statically Determinate Beams

A statically determinate structure is the one in which reactions and internal forces can be determined solely from equations of equilibrium and free-body diagrams. These equations are: $\sum H = 0$, $\sum V = 0$ and $\sum M = 0$. It should be noted that the results of analysis are independent of the material from which the structure has been fabricated.

Numerical Problems

1. Let us determine the support reactions for the beam shown in figure at A and B.

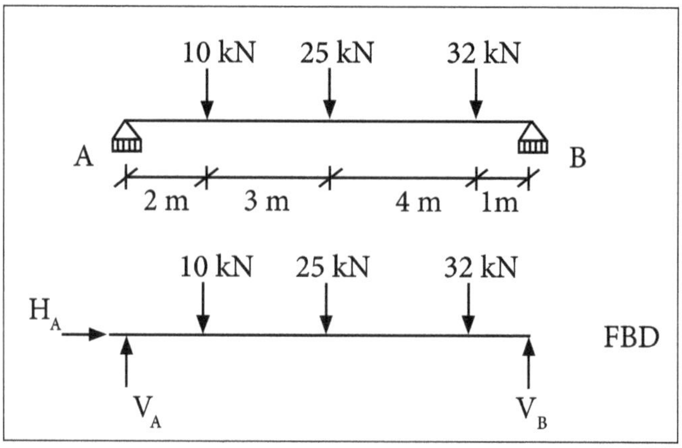

Solution:

$$\sum f_{x1} = 0$$

$$\sum f_{y1} = 0$$

$\sum M_0 = 0$

$V_A - 10 - 25 - 32 + V_B = 0$

$V_A + V_B = 67$ kN

$\zeta + \sum M_A = 0$

$-10(2) - 25(5) - 32(9) + V_B(10) = 0$

$V_B = 43.3$ kN

$V_A = 23.7$ kN

2. Let us determine the support reactions for the beam shown in figure at A and B.

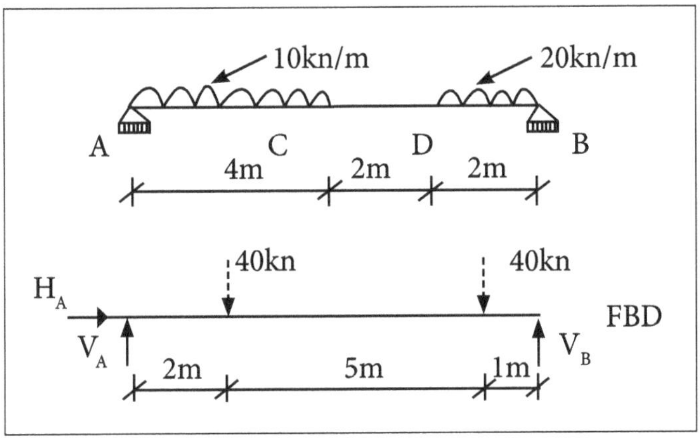

Solution:

$\sum f_{x1} = 0,\ H_A = 0$

$\sum f_{y1} = 0$

$V_A - 40 - 40 + V_B = 0$

$V_A + V_B = 80$ kN

$\zeta + \sum M_A = 0$

$0 - 40(2) - 40(7) + V_B(8) = 0$

$V_B = 45$ kN

$V_A = 35$ kN

3. Let us determine the support reactions for the beam shown in figure at A and B.

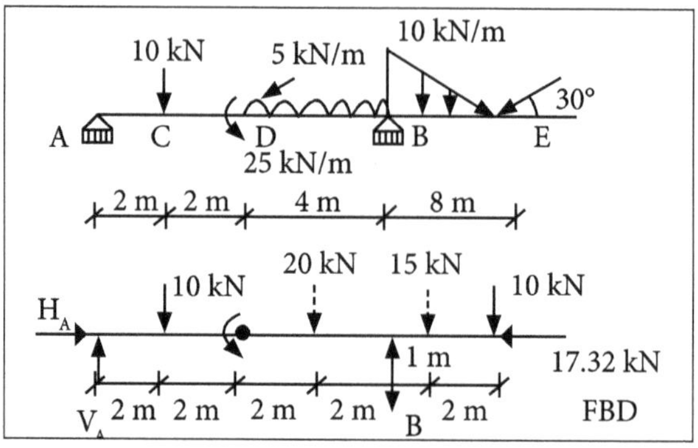

Solution:

$$\sum f_{x1} = 0,$$

$$H_A - 17.32 = 0$$

$$H_A = 17.32 \text{ kN}$$

$$\sum f_{y1} = 0$$

$$V_A - 10 - 20 - 15 - 10 + V_B = 0$$

$$V_A + V_B = 55 \text{ kN}$$

$$\zeta + \sum M_A = 0$$

$$(-10 \times 2) + 25 - 20(6) + V_B(8) - 15(9) - 10(11) = 0$$

$$V_B = 45 \text{ kN}$$

$$V_A = 10 \text{ kN}$$

4. Let us determine the support reactions for the beam shown in figure at A and B.

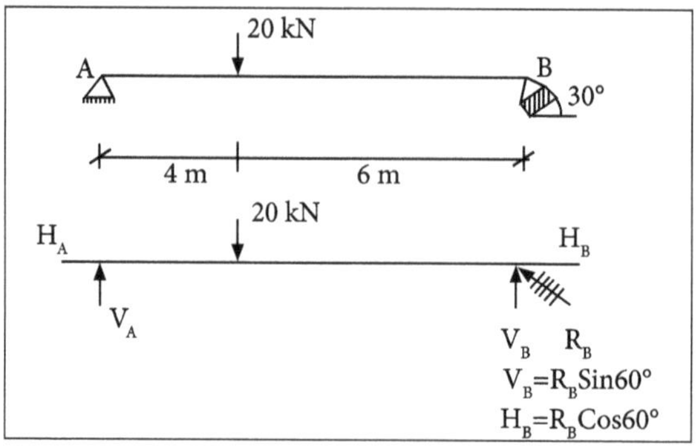

Solution:

$$\sum f_{x1} = 0,$$

$$H_A - R_B \sin 30° = 0$$

$$H_A = 0.5R_B$$

$$\sum f_{y1} = 0$$

$$V_A - 20 + R_B \cos 30° = 0$$

$$V_A + 0.866R_B = 20$$

$$\zeta + \sum M_B = 0$$

$$-V_A(10) + 20(6) = 0$$

$$-V_A = 12 \text{ kN}$$

$$R_B = 9.24 \text{ kN}$$

$$H_A = 4.62 \text{ kN}.$$

4

Centroids and Moments of Inertia of Engineering Sections

4.1 Centroids

The center of gravity of a body is defined as the point through which the entire weight of the body acts. When this is referred to weightless laminas or plane area it is called the centroid of the area.

Difference between Centroid and Centre of Mass

The centroid is purely a geometrical thing. It is the centre of gravity for objects of uniform density. But an object's density can be non-uniform, which will move the centre of gravity away from the centre. Imagine a 2D shape made from cardboard. The point in the middle where we can balance it on the end of a pin is the centroid. But if we then replace part of the shape with steel, keeping the same shape, the centroid is the same but the centre of gravity has shifted.

4.1.1 Centroid of Line and Area

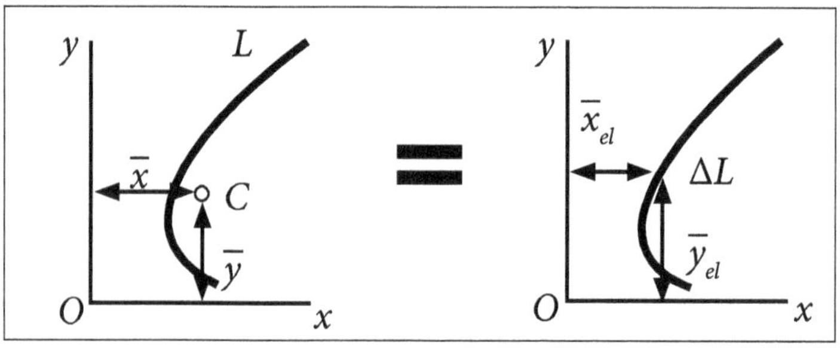

$$\bar{x}L = \int_L \bar{x}_{el} dL$$

$$\bar{x}L = \int_L \bar{y}_{el} dL$$

$$\equiv \int_A^{i=1} \bar{y}_{el} dA$$

Centroids of Area

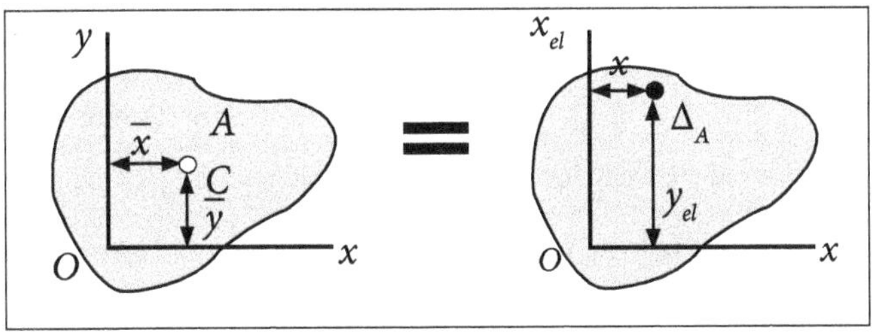

$$\bar{x}A = \sum_{i=1}^{n} \bar{x}_i A_i$$

$$= \int_A \bar{x}_{el} dA$$

$$\bar{y}A = \sum_{i=1}^{n} \bar{y}_i A_i$$

$$= \int_A \bar{y}_{el} dA$$

4.1.2 Centroid of Basic Geometrical Figures

Centroid of a Rectangular Section by Integration

Figure shows a rectangular section ABCD having width = b and depth = d. Consider a rectangular elementary strip of thickness 'dy' at a distance y from the axis OX.

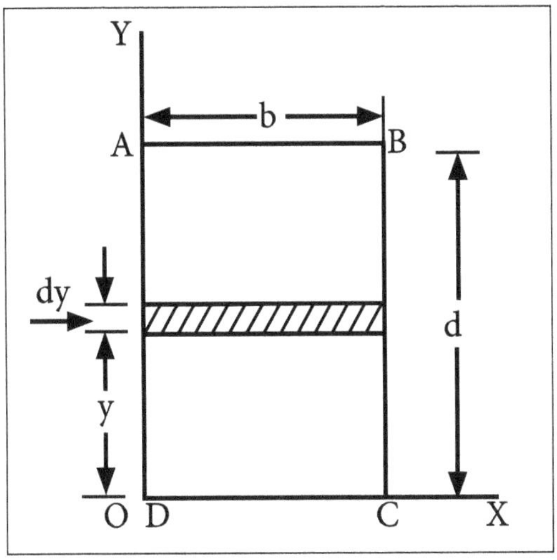

(a)

Let,

dA = Area of strip = b.dy

Moment of the area dA about axis OX

= dA x y = (b.dy) x y (∵ dA = b.dy)

=by x dy

The moment of the whole area about axis OX (or first moment of the whole area about axis OX) will be obtained by integrating the above equation between the limits o to d.

$$= \int_{o}^{d} \text{by x dy} = b\int_{o}^{d} ydy$$

(∵ b is constant and can be taken outside the integral sign)

$$= b\left[\frac{y^2}{2}\right]_o^d = \frac{bd^2}{2} \quad ...(1)$$

Let A = Total area of rectangular section,

$$\int_o^d dA = \int_o^d b\ dA = b[y]_o^d \quad [\because dA = b.dy]$$

$\bar{y}$ = Distance of the centroid of the rectangular section from axis OX.

Moment of total area of rectangular section about axis OX,

$$= A \times \bar{y} = (b \times d) \times \bar{y} \quad ...(2)$$

By equating the 1 and 2, we get,

$$(b \times d) \times \bar{y} = \frac{db^2}{2}$$

$$\therefore \bar{y} = \frac{db^2}{2} \times \frac{1}{bd} = \frac{d}{2}$$

Similarly the distance of the centroid of the rectangular section from the axis OY is,

$$\bar{x} = \frac{b}{2}$$

Refer to figure a,

Area of strip, dA = d.dx

Moment of this area dA about axis OY,

= dA × x

Moment of the whole area about axis OY,

$$= \int_{o}^{b} dA \times x = \int_{o}^{b} (d.dx)x$$

$$[\because dA = d.dx]$$

$$= d\int_{o}^{b} x\ dx = d\left[\frac{x^2}{2}\right]_o^b = \frac{d \times b^2}{2} \quad ...(3)$$

Let $\overline{x}$ = Distance of centroid of whole area from axis OY.

A = Total area = b x d

∵ Moment of total area A about axis OY,

$$= A \times \overline{x}$$

Equating equations (3) and (4), we get

$$A \times \overline{x} = \frac{b \times d^2}{2}$$

or,

$$\overline{x} = \frac{b \times d^2}{2} \times \frac{1}{A}$$

$$= \frac{b \times d^2}{2} \times \frac{1}{b \times d} = \frac{b}{2}$$

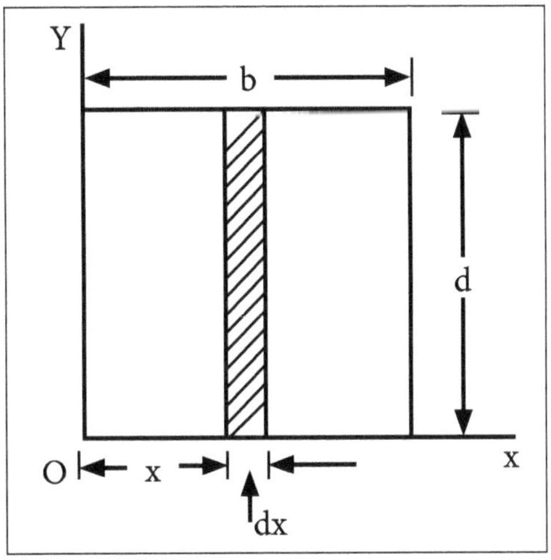

(b)

Figure 1b shows a circular section of radius R with o as centre. The equation of the circle is given by,

$$x^2 + y^2 = R^2$$

Consider a rectangular elementary strip of thickness dy at a distance of y from the axis OX.

The area of strip, dA=2x.dy.

Moment of this area dA about x-axis,

$$= dA.\ Y$$

$$= (2x.dy).y\ (\because dA = 2x.dy)$$

$$= 2xy.dy \qquad ...(1)$$

But the equation of the circle is,

$$x^2 + y^2 = R^2$$

or,

$$x = \sqrt{R^2 - y^2}$$

Substituting the above value of x in equation 1, we get moment of area dA about x-axis

$$= 2\sqrt{R^2 - y^2}.y.dy$$

Moment of total area A about x-axis will be obtained by integrating the above equation from -R to R.

Moment of area A about x-axis.

$$= \int_{-R}^{+R} 2\sqrt{R_2 - y^2}\,(-2y).dy$$

$$= -\left[\frac{\left(R^2 - y^2\right)^{3/2}}{3/2}\right]_{-R}^{R}$$

$$= \frac{2}{3}\left[\left(R^2 - R^2\right)^{3/2} - \left\{\left(R^2\left(-R^2\right)\right\}^{3/2}\right]$$

$$= \frac{-2}{3}[0 - 0] = 0 \qquad ...(2)$$

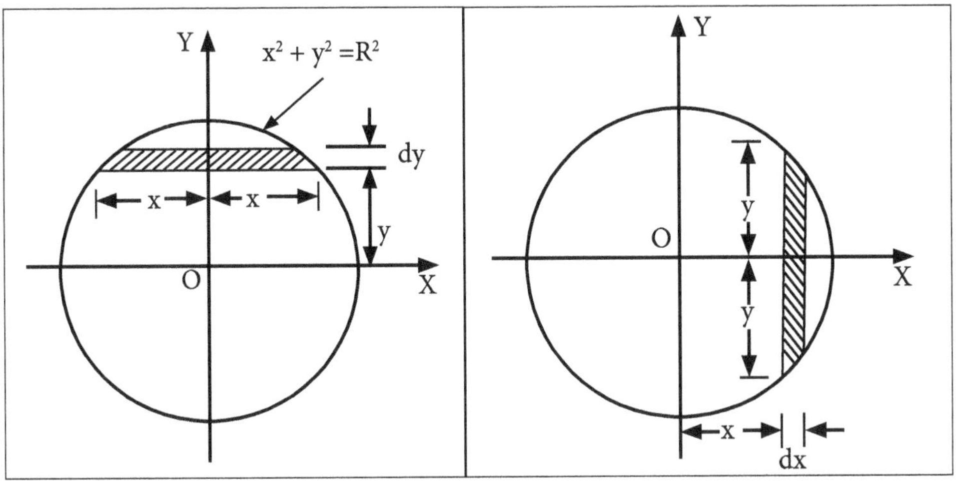

(2a) & (2b)

Also the moment of total area A about x-axis = $A \times \bar{y}$ ...(3)

Where $\bar{y}$ = Distance of centroid of total area A from x-axis.

Equating the two values given by equations 2 and 3, we get

$$A \times \bar{y} = 0 \quad \text{or} \quad \bar{y} = 0$$

This means that the centroid of the circle is at the centre of the circle.

Centroid of a Triangular Section by Integration

Figure 3 shows a triangular section AOB of base width = b and height = h.

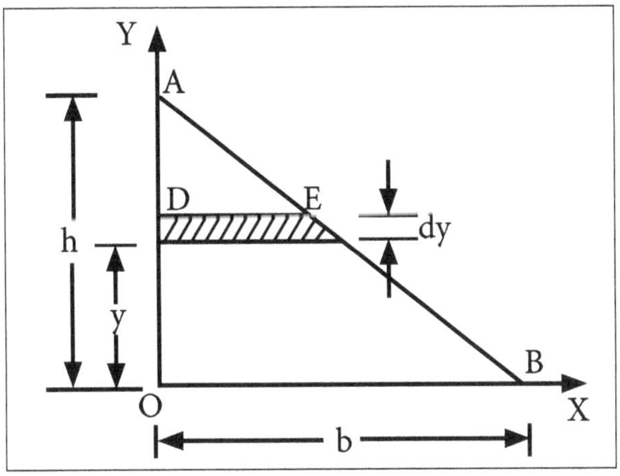

(3)

Consider a small strip of thickness dy at a distance y from the axis OX.

Area of strip, dA = Length DE $\times dy$.

The distance DE in terms of y, b and h is obtained from similar triangles ADE and AOB as,

$$\frac{DE}{OB} = \frac{AD}{AO}$$

Where OB = b, AO =h and AD = (h-y)

$$\therefore \quad \frac{DE}{b} = \frac{(h-y)}{h}$$

Or,

$$DE = \frac{b(h-y)}{h}$$

Substituting this value of DE in equation 1, we get

$$\text{Area of strip, } dA = \frac{b(h-y)}{h}.dy$$

Moment of this area dA about axis OX,

$$dA.y = \frac{b(h-y)}{h}.dy.y$$

$$\frac{b}{h}(h-y).y.dy$$

The moment of the total area A of the triangular section is obtained by integrating the above equation between the limits 0 to h.

$$= \frac{b}{h}\left[h^2 \frac{h^2}{2}\right]$$

$$= \frac{b}{h}.\frac{h^2}{2} = \frac{bh}{2}$$

Then moment of total area A about axis OX,

$$= A \times \bar{y}$$

$$= \left(\frac{b \times h}{2}\right) \times \bar{y}$$

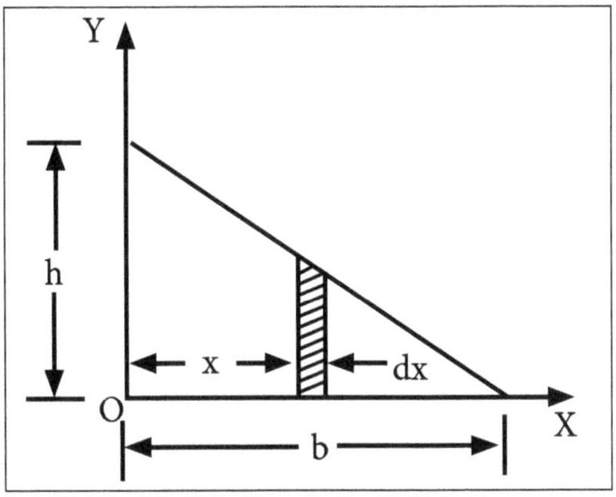

(3a)

Equating the two values given by equation (2) and (3), we get,

$$= \left(\frac{b \times h}{2}\right) \times \bar{y}$$

$$= \frac{b}{h} \cdot \frac{h^3}{6}$$

or,

$$\bar{y} = \left(\frac{b}{h} \cdot \frac{h^3}{6}\right) \times \frac{1}{\left(\frac{b \times h}{2}\right)} = \frac{h}{3}$$

4.1.3 Numerical Problems

1. Let us determine centroid of the shaded area with reference apex.

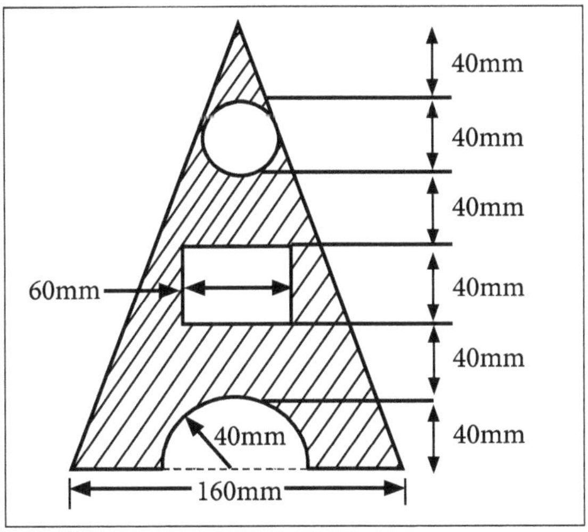

Solution:

Given:

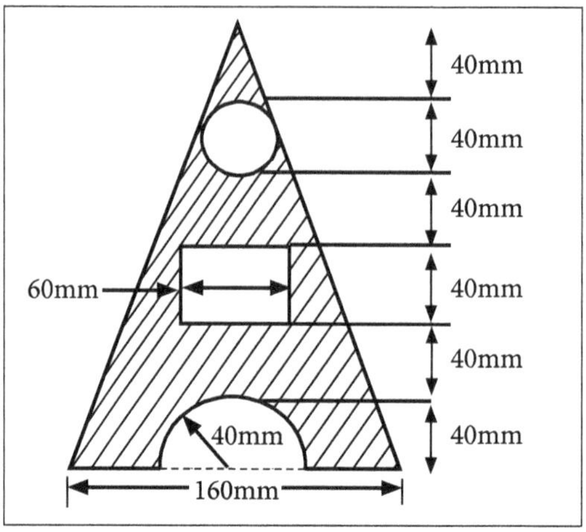

For the given figure, there is a vertical axis of symmetry. Hence we have to find $\bar{y}$ only.

Given shaded area = Triangle - Circle - Rectangle - Semicircle.

Assuming center of semicircle as origin, measure the centroidal y distances as shown in the table below:

Component area a (mm²)	Vertical (y) centroidal distances(mm)	Product a.y (mm³)
$\frac{1}{2}(160)240 = 19200$	$\frac{1}{3}(240) = 80$	+1536000
$-\frac{\pi}{4}(40)^2 = -1256.64$	$160 + \frac{40}{2} = 180$	−226195.20
$-(40)(60) = -2400$	$80 + \frac{40}{2} = 180$	−240000
$-\frac{\pi}{4}(40)^2 = -2513.27$	$\frac{4(40)}{3\pi} = 16.98$	−42675.325
$\sum A = 13030.09$	-	1027129.5

Apply $\left(\sum A\right)\bar{y} = \sum(a.y)$

$$\therefore = \bar{y}\ 78.83 \text{ mm}$$

∴ Distance from apex will be 161.17 mm

2. Let us determine the position of the centroid of the solid combination consisting of a solid cone of height 50mm and base diameter 80 mm and a cylinder of height 100mm and diameter 80 mm with a semicircular cut as shown.

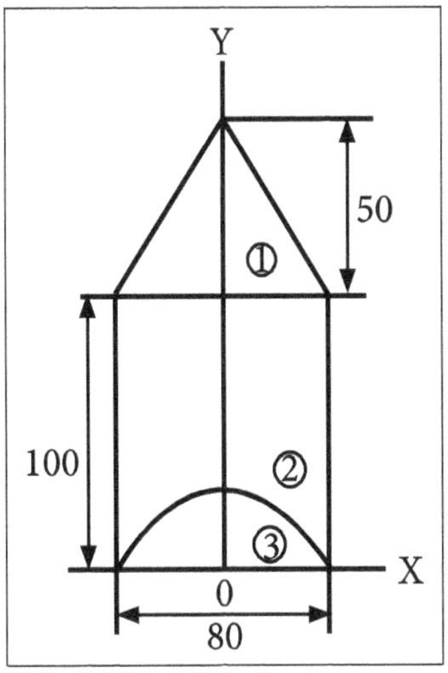

Solution:

Given:

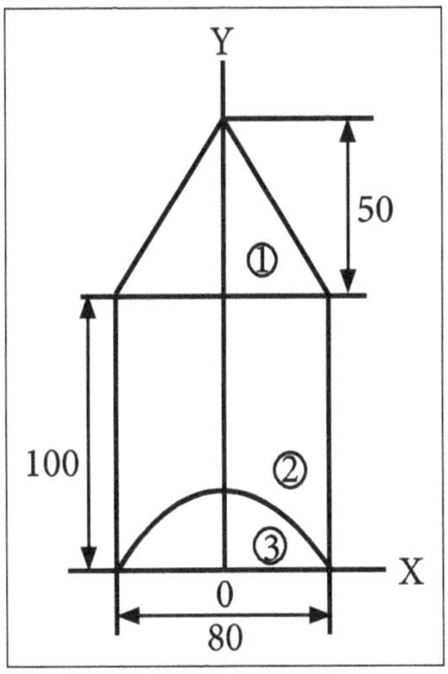

The given figure is symmetrical about both the axes.

Section (1) Triangle,

Area $A_1 = (1/2)\ bh$

$$A_1 = \frac{1}{2}(80)(50)$$

$A_1 = 2000\ mm^2$

$X_1 = (b/2) = (80/2) = 40\ mm$

$$Y_1 = (n/3) = (50/3) = 16.66 \text{ mm}$$

Section (2) Rectangle,

$$A_2 = lb$$

$$A_2 = 80 \times 100$$

$$A_2 = 8000 \text{ mm}^2$$

$$X_2 = (l/2) = (80/2) = 40 \text{ mm}$$

$$Y_2 = (b/2) = (100/2) = 50 \text{ mm}$$

Section (3) Semicirclel,

$$A = \frac{1}{2} \times \frac{\pi d^2}{4}$$

$$A_3 = \frac{1}{2} \times \frac{\pi (80)^2}{4}$$

$$A_3 = 2513.2 \text{ mm}^2$$

$$X_3 = \frac{d}{2} = \frac{80}{2} = 40\text{mm}$$

$$Y_3 = \frac{4x}{3\pi} = \frac{4(40)}{3\pi} = 167.5\text{mm}$$

$$\bar{X} = \frac{a_1 X_1 + a_2 X_2 + a_3 X_3}{a_1 + a_2 + a_3}$$

$$\bar{X} = \frac{2000(40) + 8000(40) + (2513.2)(40)}{2000 + 8000 + 2513.2}$$

$$\bar{X} = \frac{500528}{12513.2}$$

$$\bar{X} = 40\text{mm}$$

$$\bar{Y} = \frac{a_1 Y_1 + a_2 Y_2 + a_3 Y_3}{a_1 + a_2 + a_3}$$

$$\bar{Y} = \frac{2000(16.67) + 8000(50) + 2513.2(167.5)}{2000 + 8000 + 2513.2}$$

$$\bar{Y} = \frac{854301}{12513.2}$$

$$\bar{Y} = 68.27\text{mm}$$

3. Let us calculate the centroid of the shaded area OPQ shown in figure. The curve OQ is parabolic,

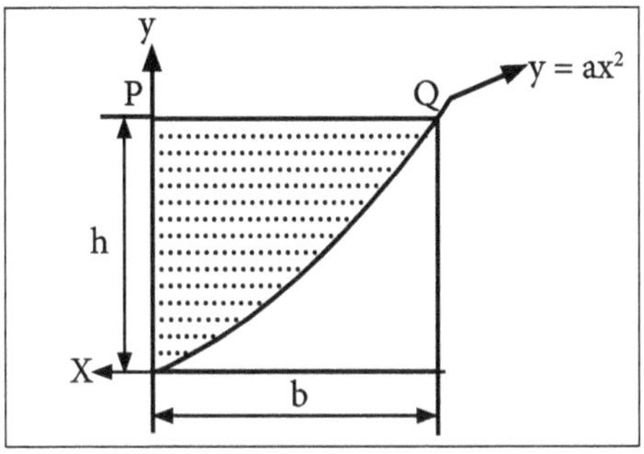

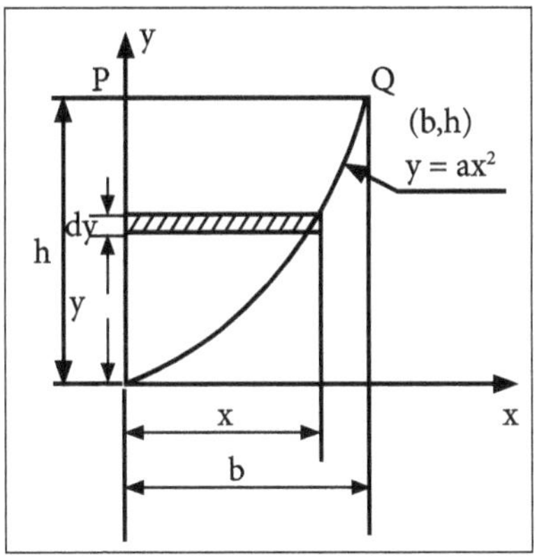

Solution:

Given:

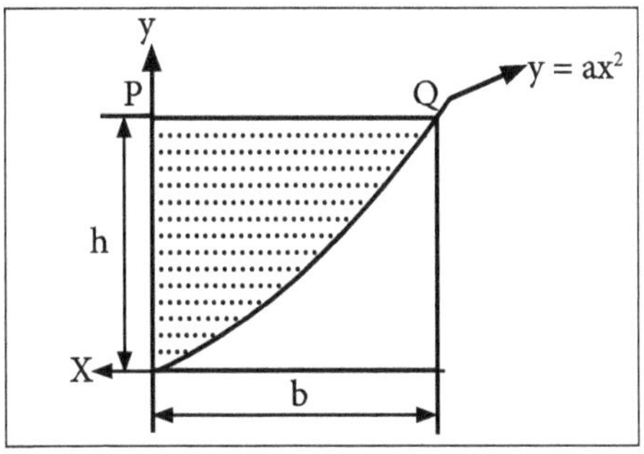

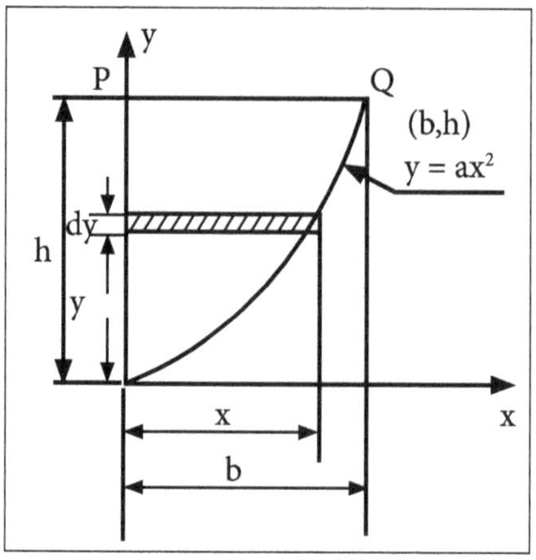

The equation of the curve is:

$$y = ax^2 \quad \text{...(1)}$$

If y = h and x = b

$$h = ab^2$$

$$a = \frac{h}{b^2}$$

$$y = ax^2$$

$$x = \left(\frac{y}{a}\right)^{1/2} \quad \text{...(2)}$$

Consider horizontal strip of area dA = x . dy,

$$\therefore A = \int_0^h x \cdot dy$$

$$A = \int_0^h \left(\frac{y}{a}\right)^{1/2} \cdot dy$$

$$A = \frac{1}{\sqrt{a}}\left[\frac{y^{3/2}}{3/2}\right]_0^h$$

$$A = \frac{2}{3\sqrt{a}} \cdot h^{3/2}$$

$$A = \frac{2}{3\left(\frac{h}{b^2}\right)^{1/2}} \cdot h^{3/2}$$

$$A = \frac{2}{3} bh$$

Moment of elemental area about X Axis is given by,

$$= y \cdot dA$$

$$= y \cdot (x\,dy)$$

Moment of entire area,

$$= \int_0^h y \cdot x \cdot dy \quad \left[\therefore = \int y \cdot dA\right]$$

$$= \int_0^h y \cdot \left(\frac{y}{a}\right)^{1/2} \cdot dy$$

$$= \frac{1}{\sqrt{a}} \int_0^h y^{3/2} \cdot dy$$

$$\frac{1}{\sqrt{}} \left[\frac{y^{5/2}}{5/2}\right]$$

$$\int y \cdot dA = \frac{2}{5\sqrt{a}} \cdot h^{5/2}$$

From that,

$$\bar{y} = \frac{\int y \cdot dA}{A}$$

$$= \frac{\frac{2}{5\sqrt{a}} \cdot h^{5/2}}{2/3 \cdot bh}$$

$$= \frac{-3}{2} \times \frac{2}{5\sqrt{a}} \frac{h^{5/2}}{b \cdot h}$$

$$y = \frac{3}{5b} \cdot \frac{h^{3/2}}{\sqrt{a}} \quad \left[\because a = \frac{h}{b^2}\right]$$

$$= \frac{3}{5b} \cdot \frac{h^{3/2}}{\;^{1/2}}$$

$$\bar{y} = \frac{3}{5}h$$

Moment of elemental area about Y-axis,

$$= \frac{x}{2} \cdot dA$$

$$= \frac{x}{2} \cdot x \cdot dy$$

Moment of entire area is given by,

$$= \int_0^h \frac{x}{2} \cdot dA$$

$$= \int_0^h \frac{x}{2} \cdot x \cdot dy$$

$$= \frac{1}{2}\int_0^h x^2 \cdot dy$$

$$= \frac{1}{2}\int_0^h \frac{y}{a} \cdot dy \quad \left[\because x = \left(\frac{y}{a}\right)^{1/2} \right]$$

$$= \frac{1}{2a}\left[\frac{y^2}{2}\right]_0^h$$

$$= \frac{1}{4} h b^2$$

$$\bar{x} = \frac{\int \frac{x}{2} \cdot dA}{A}$$

$$= \frac{\frac{1}{4} \cdot hb^2}{2/3 \cdot bh}$$

$$= \frac{1}{4} \times \frac{3}{2} \times \frac{hb^2}{bh}$$

$$\bar{X} = \frac{3}{8} b$$

Result:

$$\bar{X} = \frac{3}{8} b$$

$$\bar{Y} = \frac{3}{5} h$$

4. Let us locate the centroid of the area shown in figure below. The dimensions are in mm.

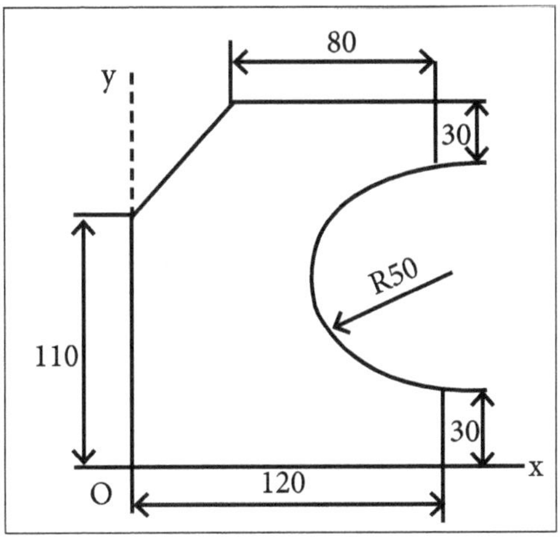

Solution:

Given:

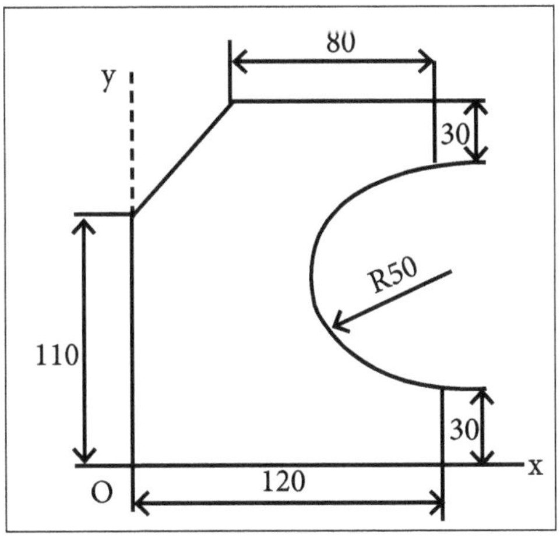

Section	Area mm²	X-Co- ordinate of C.G $(X)_{(mm)}$	Y-Co- ordinate of C.G $(Y)_{(mm)}$	A_X (mm³)	A_Y (mm³)
1) Rectangle	$A_1 = 120 \times 16 =$ 19200	$x_1 = \frac{120}{2}$ $= 60$	$y_1 = \frac{160}{2}$ $= 80$	1520,000 (+)	1536,000 (-)
2) Triangle	$A_2 = \frac{1}{2} bh$ $= \frac{1}{2} \times 40 \times 50$ $= 1000$	$x_2 = \frac{40}{2}$ $= 20$	$y_2 = 110 + \left(50 - \frac{50}{3}\right)$ $y_2 = 44.33$	20,000 (-)	44,330 (-)
3) Semicircle	$A_3 = \frac{\pi r^2}{2}$ $= \pi \frac{50^2}{2}$ $= 3925$	$x_3 = 70 + \left(4 - \frac{4r}{3\pi}\right)$ $= 70 + \left(r - \frac{4r}{3\pi}\right)$ $= 70 + \left(50 - \frac{4 \times 50}{3\pi}\right)$ $= 70 + (50 - 21)$ $= 99$	$y_3 = 30 + \frac{100}{2}$ $= 80$	388,575 (-)	314,000 (-)
				$\sum A_X =$ 11,111,425 mm³	$\sum A_Y =$ 1,177,670mm³

$$\overline{x} = \frac{A_x x_1 - A_e x_2 - A_3 x_3}{A_1 - (A_2 + A_3)}$$

$$= \frac{11520,000 - (20,000) - 388,575}{19200 - (1000 + 3925)}$$

$$\overline{x} = \frac{11,111,425}{14,275} = 778.38\text{mm}$$

$\overline{x}$ = 778.38 mm

$$\overline{y} = \frac{A_1 Y_1 - A_2 Y_2 - A_3 Y_3}{(A_1 (A_1 - A_3)}$$

$$= \frac{153,6000 - 44330 - 314000}{(19200 - (1000 + 3925)}$$

$$= \frac{1176,670}{14275}$$

$\overline{y}$ =82.49mm

Result:

Centroid $\bar{x} = 778.38$ mm

$\bar{y} = 82.49$ mm

5. Let us locate the centroid of the T-section shown in the figure.

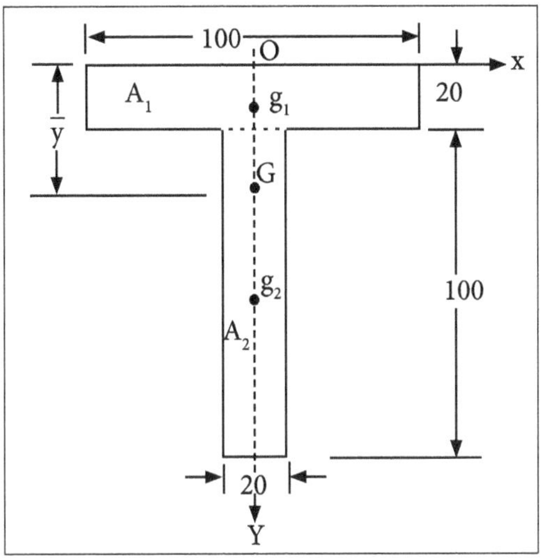

All dimensions are in mm.

Solution:

Given:

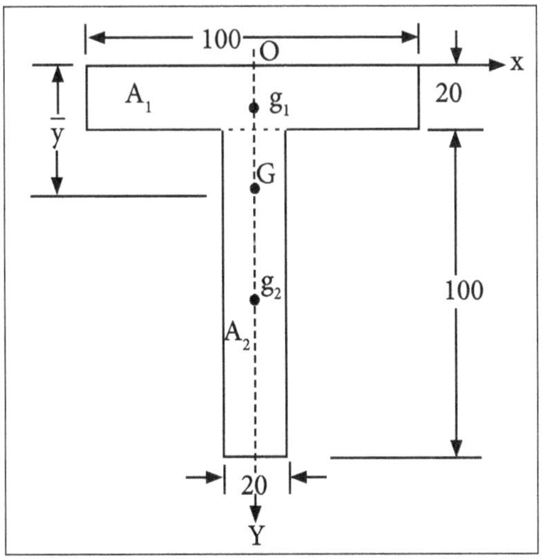

Let us select the axis as shown in figure. We can say due to symmetry centroid lies on y axis, i.e. $\bar{x} = 0$. Now the given T-section may be divided into two rectangles A_1 and A_2

each of size 100 × 20 and 20 × 100. The centroid of A_1 and A_2 are $g_1(0, 10)$ and $g_2(0, 70)$ respectively.

∴ The distance of centroid from top is given by:

$$\bar{y} = \frac{100 \times 20 \times 10 + 20 \times 100 \times 70}{100 \times 20 + 20 \times 100}$$

$$= 40\text{mm}$$

Hence, centroid of T-section is on the symmetric axis at a distance 40 mm from the top.

6. Let us locate the centroid of the I-section shown in figure.

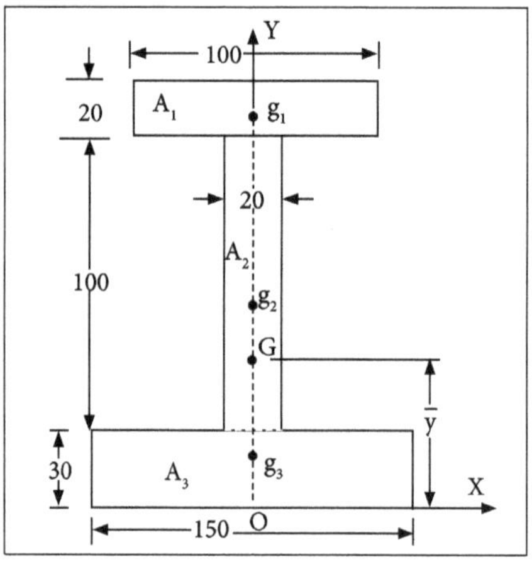

All dimensions are in mm.

Solution:

Given:

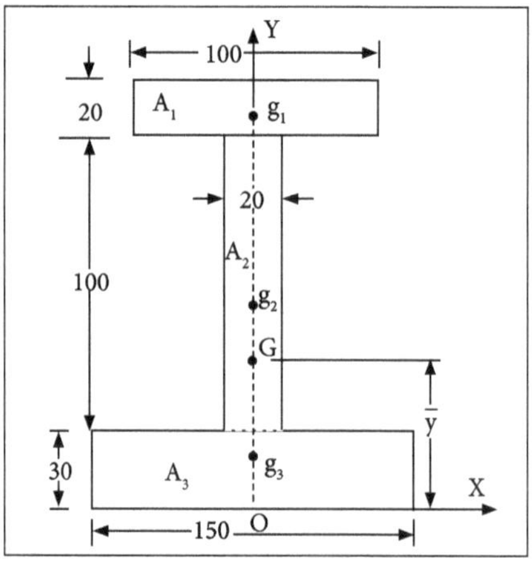

Select the co-ordinate system as shown in Figure, due to symmetry centroid must lie on y axis,

All dimensions are in mm.

i.e., $\bar{x} = 0$

Now, the composite section may be split into three rectangles,

$$A_1 = 100 \times 20 = 2000 \text{ mm}^2$$

Centroid of A_1 from the origin is,

$$y_1 = 30 + 100 + (20/2) = 140 \text{ mm}$$

Similarly,

$$A_2 = 100 \times 20 = 2000 \text{ mm}^2$$

$$Y_2 = 30 + (100/2) = 80 \text{ mm}$$

$$A_3 = 150 \times 30 = 4500 \text{ mm}^2$$

And,

$$Y_3 = (30/2) = 15 \text{ mm}$$

$$\bar{y} = \frac{A_1Y_1 - A_2Y_2 - A_3Y_3}{A}$$

$$= \frac{2000 + 140 + 2000 \times 80 + 4500 \times 15}{2000 + 2000 + 4500}$$

$$= 59.71 \text{ mm}$$

$$\therefore \bar{y} = 5971 \text{ mm}$$

Thus, the centroid is on the symmetric axis at a distance 59.71 mm from the bottom.

7. An area in the form of L section is shown in figure and let us determine the moments of inertia $I_{xx,}$ I_{yy} and I_{xy} about its centroidal axes.

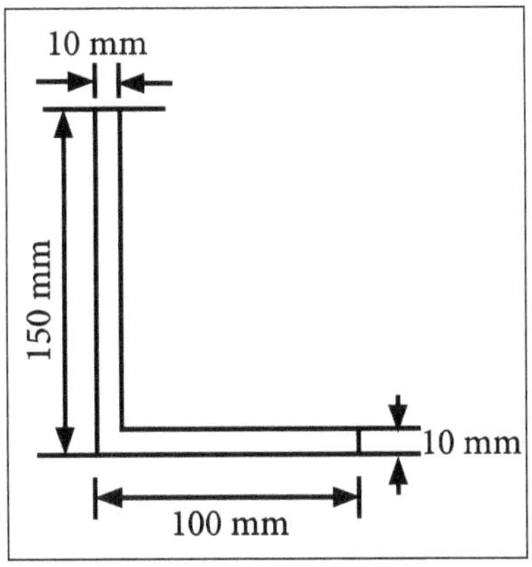

Solution:

Given:

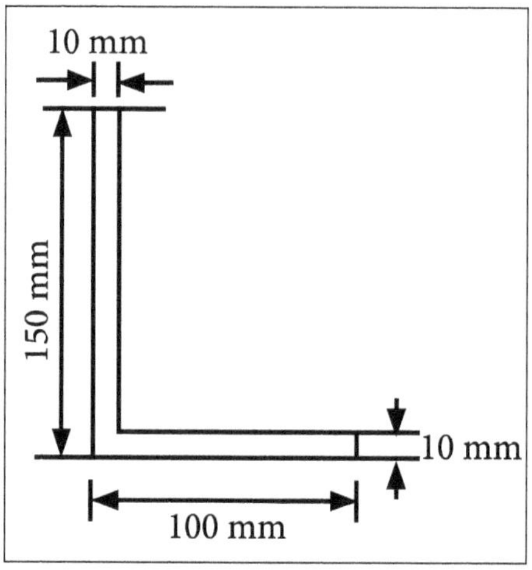

A	$\bar{x}$ $\bar{y}$	A $\bar{x}$	A $\bar{y}$	$\frac{bd^3}{12}$	Ay^2	$\frac{bd^3}{12}$	Ay^2	Axy
11000	505	50000	5000	8333	25000	833333	2500000	250000
21400	580	7000	112000	2286666	8960000	11666	35000	560000
2400		57000	117000	11279999		3379999		810000

$$\bar{x} = \frac{57000}{2400} = 23.75\text{mm}$$

$$\bar{y} = \frac{1117000}{2400} = 48.75\text{mm}$$

$$l_{xx} = \frac{bd^3}{12} - A\bar{y} \times \bar{y}$$

$$= 11279999 - 117000 \times 48.75$$

$$= 5576249 \text{ mm}^4$$

$$I_{yy} = 3379999 - 57000 \times 23.75 \Rightarrow 2026249 \text{ mm}^4$$

$$I_{yy} = 810000 - 2400 \times 23.75 \times 48.75]$$

$$I_{xy} = -1968750 \text{ mm}^4$$

4.2 Moment of Inertia

The term Moment of Inertia (I) is used to describe the capacity of a cross-section to resist bending. It is always considered with respect to the reference axis such as Y-Y or X-X. It is a mathematical property of a section concerned with the surface area and how that area is distributed about the reference axis. The reference axis is usually a centroidal axis.

The moment of inertia is also called the second moment of area and is expressed mathematically as,

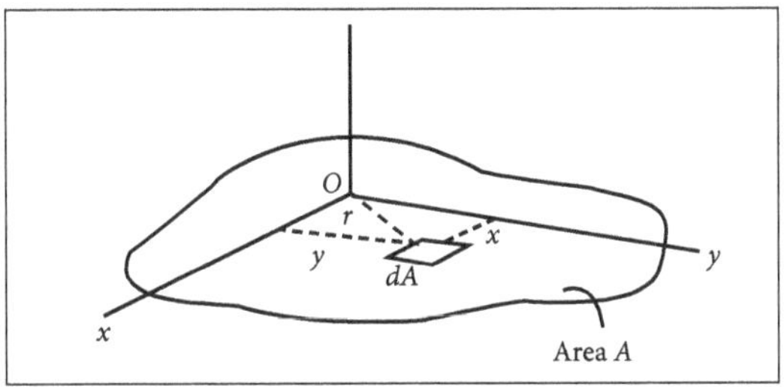

$$I_x = \int_A y^2 dA$$
$$I_y = \int_A x^2 dA$$

Where,

x = Distance from the y axis to area dA

y = Distance from the x axis to area dA

4.2.1 Radius of Gyration

The radius of gyration of an area with respect to a particular axis is defined as the square root of the quotient, of the moment of inertia divided by the area.

It is defined as the distance at which the entire area must be assumed to be concentrated in order that, the product of the area and the square of this distance will equal the moment of inertia of the actual area about the given axis.

In other words, it is defined as the way in which the total cross-sectional area is distributed around its centroidal axis. If more area is distributed further from the axis, it will have greater resistance to buckling.

Circular pipe is the most efficient column section to resist buckling, because it has its area distributed as far as possible from the centroid.

Rearranging we have,

$$I_x = k_x^2 A$$

$$I_y = k_y^2 A$$

4.2.2 Parallel Axis Theorem

The moment of inertia about different axis may be calculated using the Parallel Axis theorem, which may be written as,

$$I_{xx} = I_{cc} + Ad_{c-x}^2$$

This says that the moment of inertia about any axis parallel to an axis through the centroid of the object is equivalent to the moment of inertia about the axis passing through the centroid plus the product of the area of the object and the distance between the two parallel axis.

Radius of Gyration

$$r_{xx} = (I_{xx} / A)^{1/2}$$

The radius of gyration is the distance from an axis in which, if the entire area of the object were located at that distance, it would result in the same moment of inertia about the axis that the object has.

Polar moment of inertia,

$$J = \sum r^2 dA$$

The polar moment of inertia is the sum of the product of each bit of area dA and the radial distance to an origin squared. In a case as shown below the polar moment of inertia in related to the x & y moments of inertia by,

$$J = I_{xx} + I_{yy}$$

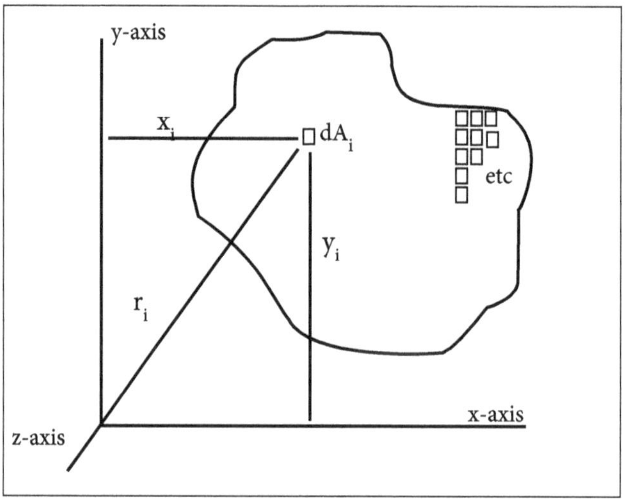

All the summations shown above become integrations if we let the dM's and dA's approach zero and while this is useful and important when calculating centroids and moments of inertia, the summation method is just as useful for understanding the concepts involved.

4.2.3 Perpendicular Axis Theorem

This theorem is applicable to the plane laminar bodies only. This theorem states that, the moment of inertia of a plane laminar about an axis perpendicular to its plane is equivalent to the sum of the moment of inertia of the lamina about two axes mutually perpendicular to each other in its plane and intersecting each other at the point where perpendicular axis passes through it.

Consider plane laminar body of arbitrary shape lying in the x-y plane as shown below in the figure:

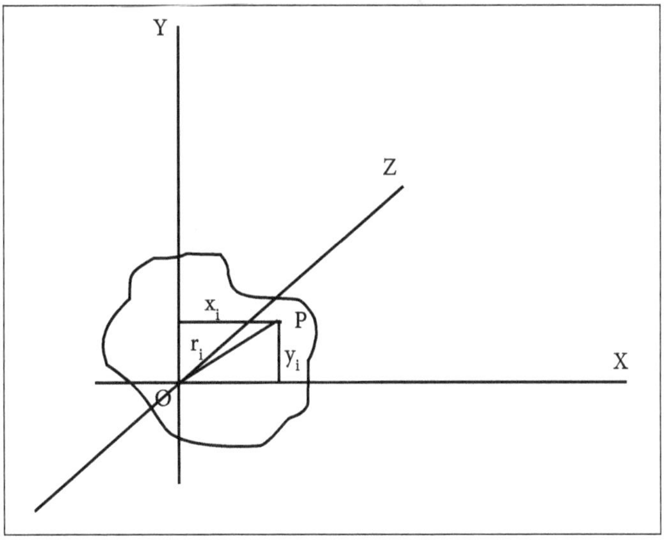

Plane laminar body with z-axis perpendicular to the plane.

The moment of inertia about the z-axis equals the sum of the moments of inertia about the x and y-axis. To prove it consider the moment of inertia about x-axis. $I_x = \sum_i m_i x_i^2$ Where, sum is taken over all the element of the mass m_i.

The moment of inertia about y axis is, $I_y = \sum_i m_i y_i^2$

Moment of inertia about z axis is, $I_z = \sum_i m_i r_i^2$ where, r_i is the perpendicular distance of particle at point P from the OZ axis. For each element, $r_i^2 = x_i^2 + y_i^2$

$$I_z = \sum_i m_i r_i^2 = \sum_i m_i \left(x_i^2 + y_i^2\right) = \sum_i m_i x_i^2 + \sum_i m_i y_i^2 = I_x + I_y$$

4.2.4 Moment of Inertia of Basic Planar Figures

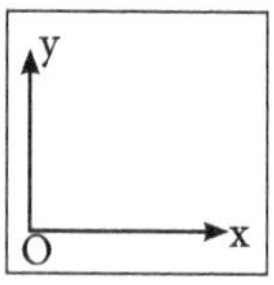	Area = A	$C_x C_y$	I_{xx}	I_{yy}
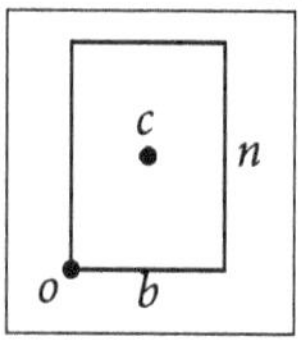 Rectangle	b.h	b/2 h/2	$A.h^2/12$	$A.b^2/12$
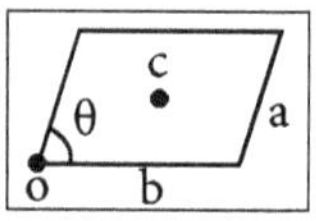 Parallelogram	a.b.sinθ	$(b+a.\cos\theta)/2 a.\sin\theta/2$	$A.(a\sin\theta)^2/12$	$A.(b^2+a^2\cos^2\theta)/12$
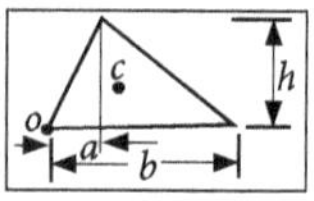 Triangle	(b.h)/2	(a +b)/3 h/3	$A.h^2/18$	$A.(b^2-a.b+a.^2)/18$
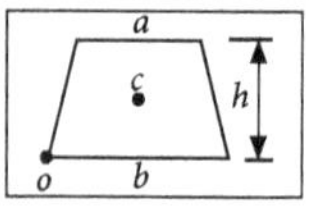 Trapezium	h.(a + b)/2	$C_y=\dfrac{h(2a+b)}{3(a+b)}$	$\dfrac{A.h^2.(a^2+4.a.b+b^2)}{18.(a+b)^2}$	-
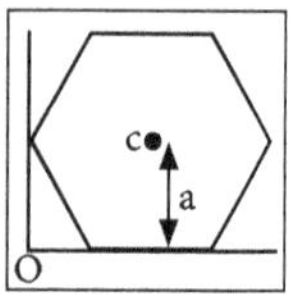 Hexagon	$6.a^2.\tan(30^\circ)$ $=3{,}464\ a^2$	a /cos(30°) = 1.155 aa	$\dfrac{A}{12}\left(\dfrac{a^2(1+2.\cos^2(30^\circ))}{\cos^2(30^\circ)}\right)$	$\dfrac{A}{12}\left(\dfrac{a^2(1+2.\cos^2(30^\circ))}{\cos^2(30^\circ)}\right)$ $=0.962\ a^4.$
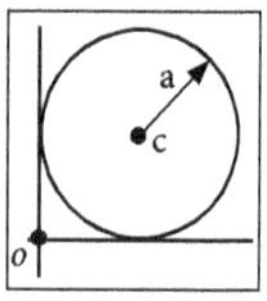 Circle	$\pi.a^2$	aa	$A.a^2/4$	$A.a^2/4$
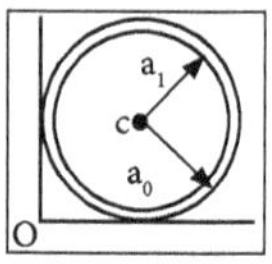 Annulus	$\pi.(a_o^2-a_i^2)$	$a_o a_o$	$\pi.(a_o^4-a_i^4)/4$	$\pi.(a_o^4-a_i^4)/4$
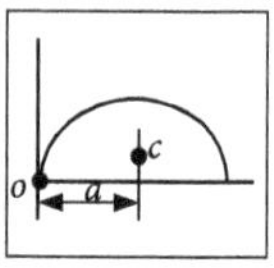 Semicircle	$\pi.a^2/2$	a4.a /3.π	$A.a^2(9\pi^2-64)/36.\pi^2$ $=0{,}1098a^4$	$A.a^2/8$

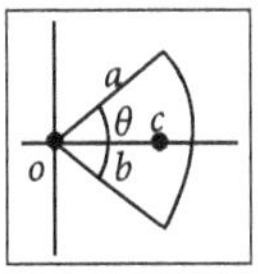 Sector of circle	$a^2.\theta$	$2.a.\sin\theta / 3.\theta o$	$\frac{A.a^2}{4\theta}(\theta-\sin(\theta)\cos(\theta))$	$\frac{A.a^2}{4\theta}\left(\theta-\sin(\theta)\cos(\theta) - \frac{16\sin^2(\theta)}{9\theta}\right)$
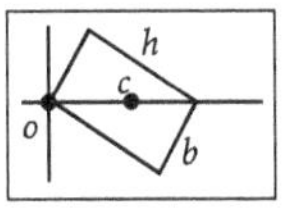 Rectangle	b.h	Sqrt($b^2 + h^2$) /20/	$A.b^2.h^2/6.(h^2+b^2)$	$A.(h^4+b^4)/12.(h^2+b^2)$
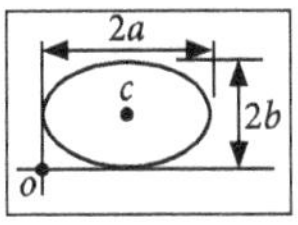 Ellipse	π.a.b	a b	$A.b^2/4$	$A.a^2/4$
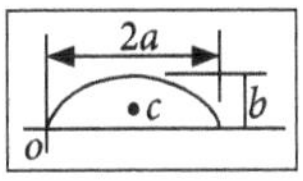 Semi ellipse	π.a.b /2	a 4b / 3π	$A.b^2(9.\pi^2-64)/36\,\pi^2$	$A.a^2/4$

4.2.5 Computing Moment of Inertia for – T, L, I, Z and Full/Quadrant Circular Sections

T Section

Section	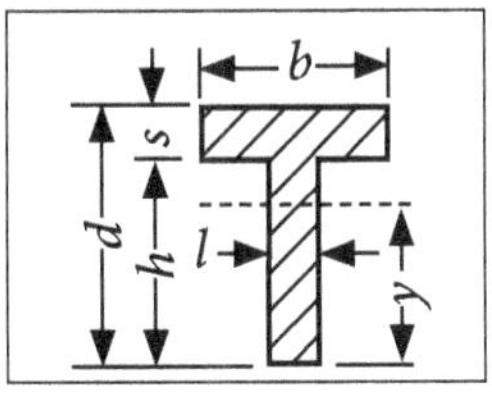	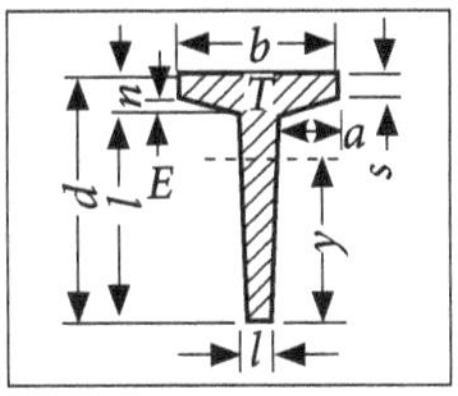	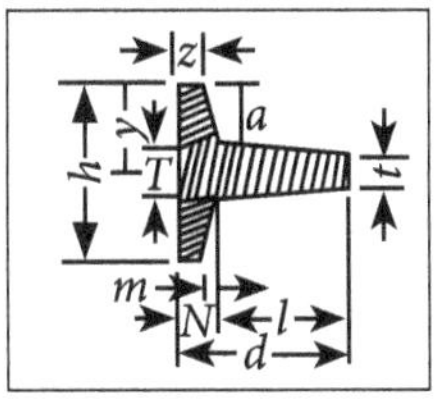
Moment of Inertia	$\frac{1}{3}\left[ty^3+b(d-y)^3 -(b-t)(d-y-s)^3\right]$	$1/12\left[l^3(T+3t)+4bn^3 -2am^3\right]$ $A(d\ \ y-n)^2$	$\frac{zb^3+mT^3-lt^3}{12} - \frac{am\left[2a^2-(2a+3T)^2\right]}{36} - \frac{l(T-t)\left[(T-t)^2-2(T+2t)^2\right]}{144}$
Radius of gyration	$\sqrt{\frac{1}{3(bs+ht)}\left[ty^3+b(d-y)^3 -(b-t)(d-y-s)^3\right]}$	$\sqrt{\frac{I}{A}}$	$\sqrt{\frac{I}{A}}$
Area	bs + ht	$I\frac{(T+t)}{2}+Tn +a(s+n)$	$\frac{I(T+t)}{2}+Tn +a(s+n)$

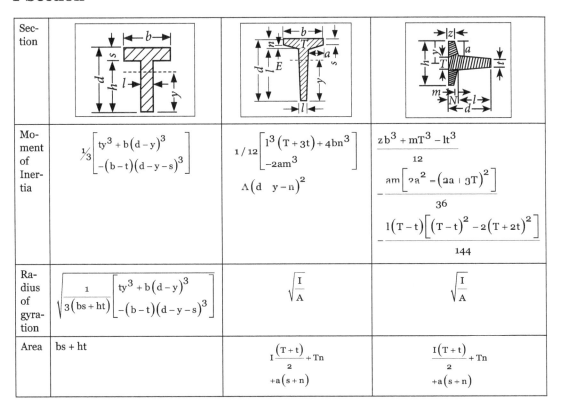

y	$d-\dfrac{d^2t+s^2(b-t)}{2(bs+ht)}$	$d-\begin{bmatrix}3s^2(b-T)+2am(m+3s)\\+3Td^2-1(T-t)(3d-1)\end{bmatrix}+6A$	$b/2$

Problems

1. Let us determine the moment of inertia of the T-section shown in figure, with respect to its centroidal X_o axis.

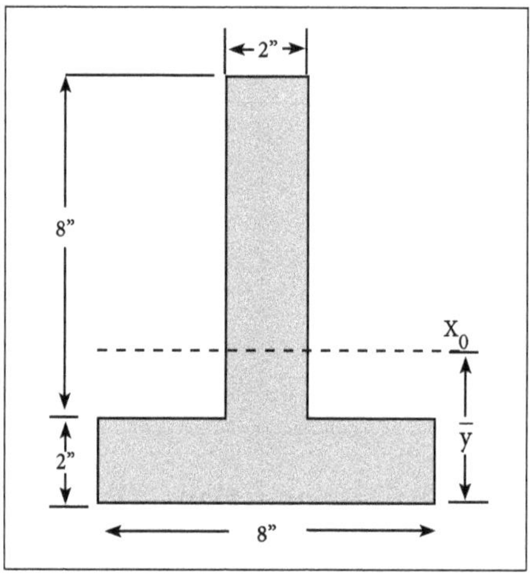

Solution:

Given:

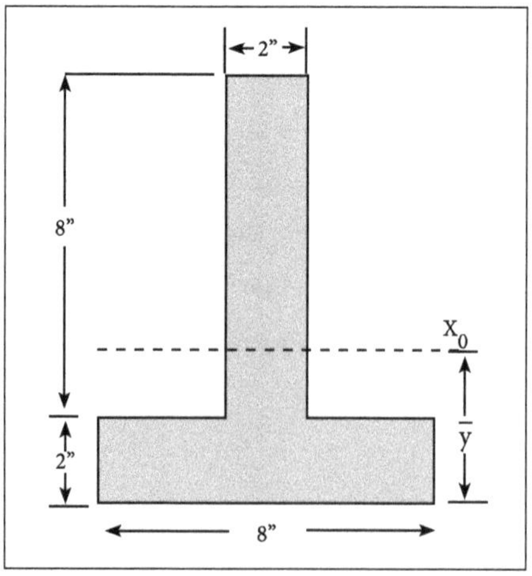

Formula to be used:

$$A\bar{y}=A_1y_1+A_2y_2$$

$$\bar{I}=\left[\bar{I}_1+A_1(\bar{y}-y_1)^2\right]+\left[\bar{I}_2+A_2(y_2-\bar{y})^2\right]$$

$$\bar{I}_1 = \frac{8(2^3)}{12} = \frac{16}{3}\text{in.}^4$$

$$A_1 = 8(2) = 16 \text{ in.}^2$$

$$y_1 = 1 \text{ in}.$$

$$A_2 = 8(2) = 16 \text{ in.}^2$$
$$y_2 = 2 + 4 = 6 \text{ in.}$$
$$A = A_1 + A_2 = 16 + 16$$
$$A = 32 \text{ in.}^2$$

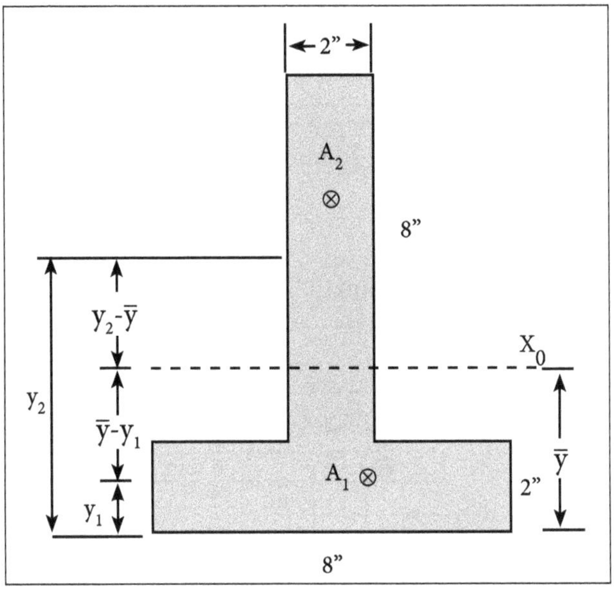

$$A\bar{y} = A_1 y_1 + A_2 y_2$$

$$32\bar{y} = 16 + 16(6)$$

$$\bar{y} = 3.5 \text{in.}$$

$$\bar{I} = \left[\frac{16}{3} + 16(3.5-1)^2\right] + \left[\frac{256}{3} + 16(6-3.5)^2\right]$$

$$\bar{I} = \left[\bar{I}_1 + A_1(\bar{y}-y_1)^2\right] + \left[\bar{I}_2 + A_2(y_2-\bar{y})^2\right]$$

$$\bar{I} = 290.67 \text{ in.}^4$$

2. Let us determine the moments of inertia and the radius of gyration of the shaded area with respect to the x and y axes.

Solution:

Given:

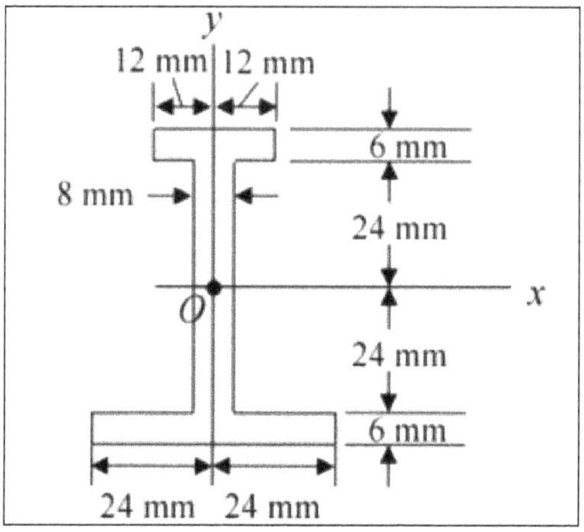

To find:

Moment of inertia and radius of gyration

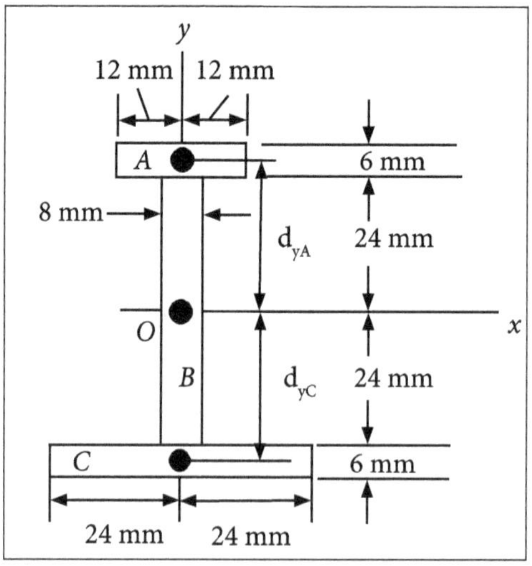

$$I_x = \left(\bar{I}_x + Ad_y^2\right)_A + \left(\bar{I}_x + A\not{d}^{\,0}{}_y^2\right)_B + \left(\bar{I}_x + Ad_y^2\right)_C$$

$$= \left[\frac{1}{12}(24)(6)^3 + (24\times6)(27)^2\right]_A + \left[\frac{1}{12}(8)(48)^3 + 0\right]_B + \left[\frac{1}{12}(48)(6)^3 + (48\times6)(27)^2\right]_C$$

$$I \quad 390\times10^3\ \text{mm}^4$$

$$k_x = \sqrt{\frac{I_x}{A}} = \sqrt{\frac{390\times10^3}{[(24\times6)+(8\times48)+(48\times6)]}} = 21.9\text{mm}$$

$$I_y = (\bar{I}_x + Ad_x^2)_A + (\bar{I}_y + Ad_x^2)_B + (\bar{I}_y + Ad_x^2)_C$$

$$\left[\frac{1}{12}(6)(24)^3\right]_A + \left[\frac{1}{12}(48)(8)^3\right]_B + \left[\frac{1}{12}(6)(48)^3\right]_C$$

$$I_y = 64.3\times10^3\ \text{mm}^4$$

$$k_y = \sqrt{\frac{L_y}{A}} = \sqrt{\frac{64.3\times10^3}{[(24\times6)+(8\times48)+(48\times6)]}} = 8.87\text{mm}$$

3. Let us determine the principal moments of inertia of the section shown in figure.

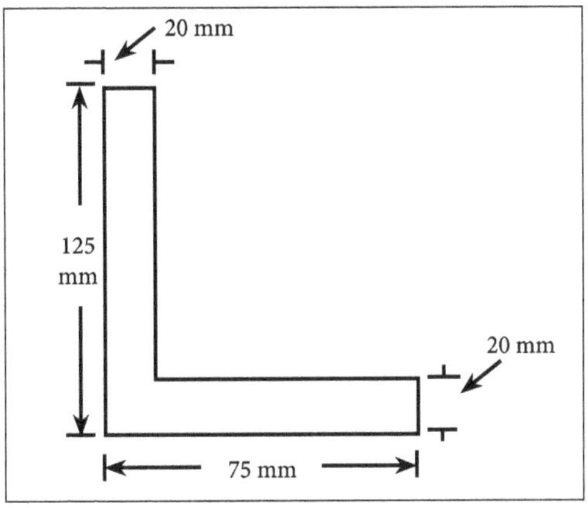

Solution:

Given:

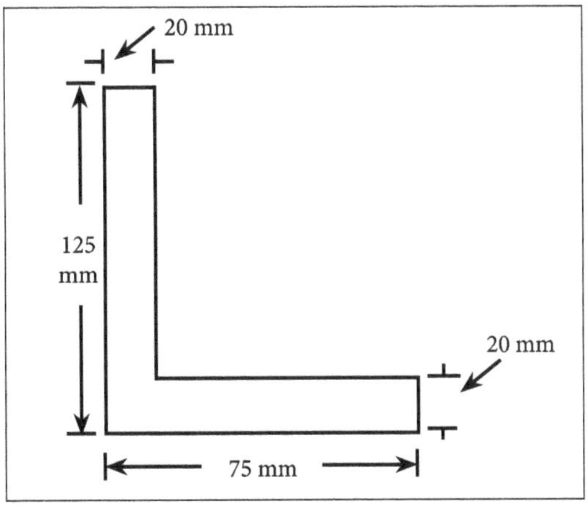

To find:

Moment of inertia

L-section: 300 mm × 20 mm × 20 mm

$I_{max} = ?\ I_{min} = ?$

For section-1

$a_1 = 280 \times 20 = 5600\ mm^2$

$x_1 = 10\ mm$

$y_1 = 20 + 140 = 160\ mm.$

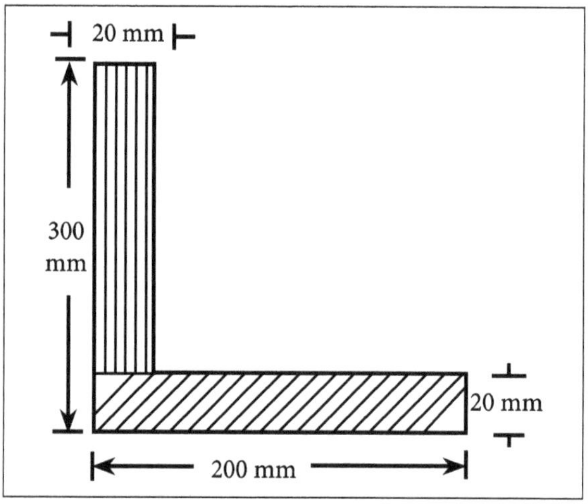

For section-2

$a_2 = 20 \times 200 = 4000\ mm^2$

$x_2 = 100\ mm$

$y_2 = 10\ mm.$

$$\bar{X} = \frac{a_1 x_1 + a_2 x_2}{a_1 + a_2}$$

$$\bar{X} = \frac{(5600 \times 10) + (4000 \times 100)}{5600 + 4000}$$

$$\bar{X} = \frac{456000}{9600}$$

$$\overline{X} = 47.5\text{mm}$$

$$\overline{Y} = \frac{a_1Y_1 + a_2Y_2}{a_2 + a_1}$$

$$\overline{Y} = \frac{(5600\times160)+(4000+10)}{5600+4000}$$

$$\overline{Y} = 97.5 \text{ mm}$$

Moment of inertia

$$I_{XX} = I_1 I_2$$

$$I_1 = (I_G)_1 + a_1h_1^2 \left\{ I_{G1} = \frac{bd^3}{12};\ h_1 = \left(\overline{Y} \approx Y_1\right) \right\}$$

$$I_1 = \frac{20\times280^3}{12} + 5600(160-97.5)^2$$

$$I_1 = 36.587\times10^6 + 21.875\times10^6$$

$$I_1 = 58.462\times10^6 \text{ mm}^4$$

$$I_2 = (I_G)_2 + a_2h_2^2 = \frac{200\times(120)^3}{12} + 4000(97.5-10)^2$$

$$I\ = 0.133\times10^6 + 30.625\times10^6$$

$$I_2 = 30.758\times10^6 \text{ mm}^4$$

$$I_{xx} = 58.462\times10^6 + 30.758\times10^6$$

$$I_{xx} - 89.22\times10^6 \text{ mm}^4$$

$$I_{YY} = I_1 + I_2 \left\{ I_{G1} = \frac{db^3}{12}; h_1 = \left(\overline{X} \approx X_1\right) \right\}$$

$$I_1 = (IG)_1 + a_1h_1{}^2$$

$$I_1 = \frac{280\times(20)^3}{12} + 5600(47.5-10)^2$$

$$I_1 = 0.1867 \times 10^6 + 7.875 \times 10^6$$

$$I_1 = 8.0617 \times 10^6 \text{ mm}^4$$

$$I_2 = (I_G)_2 + a_2 h_2^{\,2}$$

$$I_2 = \frac{20 \times (2000)^3}{12} + 4000(100 - 47.5)^2$$

$$I_2 = 13.33 \times 10^6 + 11.025 \times 10^6$$

$$I_2 = 24.355 \times 10^6 \text{ mm}^4$$

$$I_{YY} = 8.0617 \times 10^6 + 24.355 \times 10^6$$

$$I_{YY} = 32.4167 \times 10^6 \text{ mm}^4$$

Product of inertia co-ordinates,

For section-1

$$a_1 = 280 \times 20 = 5600 \text{ mm}^2$$

$$X'_1 = -(\overline{X} \approx X_1) = -(47.5 \approx 10) = -37.5 \text{mm}$$

$$Y'_1 = -(Y \approx \overline{Y}_1) = (160 - 97.5) = 62.5 \text{mm}$$

For section-2

$$a_2 = 20 \times 200 = 4000 \text{ mm}^2$$

$$X'_2 = -(\overline{X} - X_2) = (47.5 - 10) = -52.5 \text{mm}$$

$$Y'_2 = -(Y - \overline{Y}_1) = -(10 - 97.5) = 87.5 \text{mm}$$

Product of inertia,

$$I_{XY} = \sum a x' Y' = a_1 X'_1 Y'_1 + a_2 X'_2 Y'_2$$

$$I_{XY} = [5600(-37.5)(62.5)] + [4000(52.5)(-87.5)]$$

$$I_{XY} = -13.125\times10^6 - 18.375\times10^6$$

$$I_{XY} = -\ 31.5\times10^6\ mm^4.$$

Principle moment of inertia,

$$I_{max} = \left[\frac{I_{xx}+I_{YY}}{2}\right] + \sqrt{\left[\frac{I_{XX}-I_{YY}}{2}\right]^2 + I_{XY}^2}$$

$$= \left[\frac{89.22\times10^6 + 322.4167\times10^6}{2}\right] + \sqrt{\left(\frac{89.22\times10^6 - 32.4167\times10^6}{2}\right)^2 + \left(-31.5\times10^6\right)^2}$$

$$I_{max} = \left(60.818\times10^6\right) + \sqrt{\left(806.65\times10^{12}\right) + \left(992.25\times10^{12}\right)}$$

$$I_{max} = \left(60.818\times10^6\right) + \left(42.413\times10^6\right)$$

$$I_{max} = 103.23\times10^6 mm^4$$

$$I_{min} = \left[\frac{I_{xx}+I_{YY}}{2}\right] + \sqrt{\left[\frac{I_{XX}-I_{YY}}{2}\right]^2 + I_{XY}^2}$$

$$I_{min} = \left(60.818\times10^6\right) - \left(42.413\times10^6\right)$$

$$I_{min} = 18405\times10^6\ mm^4$$

To check the values:

$$I_{min} = 18405\times10^6\ mm^4$$

$$89.22\times10^6 + 23.4167\times10^6 = 103.23\times10^6 + 18.405\times lo^6$$

$$121.6367\times10^6 = 121.635\times10^6.$$

Hence, the values are correct.

Answer:

$$I_{max} = 103.23\times10^6\ mm^4$$
$$I_{min} = 18.405\times10^6\ mm^4.$$

4. Let us determine the moment of inertia of the shaded area shown in figure below about the vertical and horizontal centroidal axes. The width of the hole is 200 mm.

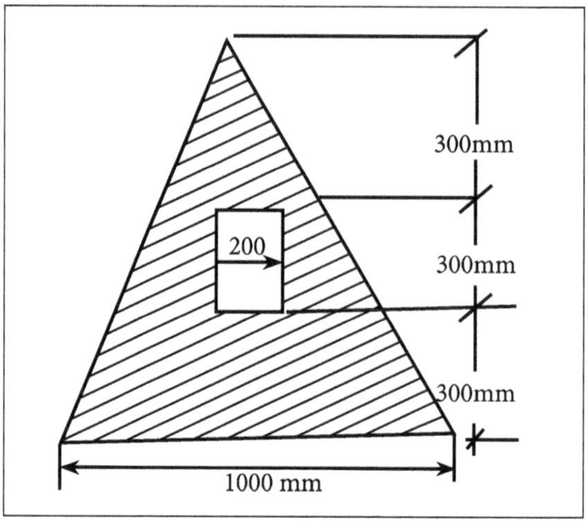

Solution:

To find:

Moment of inertia

Portion (1) Triangle:

Area $a_1 = (1/2).b.h$

$$a_1 = (1/2) \times 1000 \times 900$$

$$a_1 = 450000 = 4.5 \times 10^5 \text{ mm}^2$$

$$x_1 = (1000/2) = 500 \text{ mm}$$

$$y_1 = (900/2) = 450 \text{ mm}$$

Portion (2) Rectangle,

Area $a_2 = L \times b$

$$a_2 = 200 + 300$$

$$a_2 = 60000 \text{ mm}^2$$

$$x_2 = (200/2) = 100 \text{ mm}$$

$$y_2 = 300 + (300/2) = 450 \text{ mm}$$

$$\bar{x} = \frac{a_1x_1 - a_2x_2}{a_1 - a_2}$$

$$\bar{x} = \frac{4.5\times10^5\times500 - 6000\times100}{450000 - 60000}$$

$$\bar{x} = 561.5\,\text{mm}$$

$$\bar{y} = \frac{a_1y_1 - a_2y_2}{a_1 - a_2}$$

$$\bar{y} = \frac{\left(4.5\times10^5\times450\right) - \left(60000\times450\right)}{450000 - 60000}$$

$$\bar{y} = 450\ \text{mm}$$

Moment of inertia about XX-axis $I_{xx} = I_1 - I_2$

$$I_1 = I_{G1} + A\bar{h}_1$$

$$= \frac{1000\times(900)^3}{36} + \left[\frac{1}{2}1000\times900\times0\right]$$

$$I_1 = 2.025\times10^{10}\ \text{mm}^4$$

$$I_2 = I_{G2} + A\bar{h}_2$$

$$= \frac{(200)^4}{15} + \left[200\times300\times10\right]$$

$$I_0 = 1.33\times10^8\ \text{mm}^4$$

$$I_{xx} = I_1 - I_2$$

$$I_{xx} = 2.025\times10^{10} - 1.33\times10^8$$

$$I_{xx} = 2.0117\times10^{10}\ \text{mm}^4$$

Moment of inertia about YY-axis $I_{yy} = I_1 - I_2$

$$I_1 = I_{G1} + A\bar{h}$$

$$I_1 = \frac{h\,b^3}{48} + \left[\frac{1}{2}1000 \times 900 \times 61.5\right]$$

$$I_1 = \frac{900 \times (1000)^3}{48} + \left[27.67 \times 10^6\right]$$

$$I_1 = 1.87510 + 27.67 \times 10^6$$

$$I_1 = 1.877 \times 10^{10}\ \text{mm}^4$$

$$I_2 = I_{G2} + A\bar{h}1$$

$$I_2 = \frac{(200)^4}{12}\left[200 \times 300 \times 461.5\right]$$

$$I_2 = 1.61 \times 10^8\ \text{mm}^4$$

$$I_{yy} = I_1 - I_2$$

$$I_{yy} = 1.877 \times 10^{10} - 1.61 \times 10^8$$

$$I_{yy} = 1.869 \times 10^{10}\ \text{mm}^4.$$

5

Kinematics

5.1 Displacement

The displacement is defined as the shortest distance from the initial to the final position of a point P. Thus, it is the length of an imaginary straight path, which is typically distinct from the path actually traveled by P. A displacement vector represents the direction and length of this imaginary straight path.

A position vector expresses the position of a point P in space in terms of displacement from an arbitrary reference point O. It indicates both the distance and direction of an imaginary motion along a straight line from the reference position to the actual position of the point.

A displacement may also be defined as the 'relative position': the final position of a point (R_f) relative to its initial position (R_i) and the displacement vector can be mathematically defined as the difference between the final and initial position vectors.

$$s = R_f - R_i = \Delta R$$

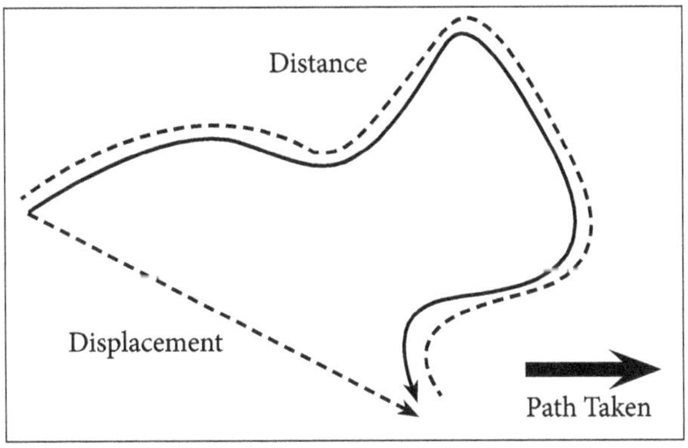

In considering the motions of objects over time, the instantaneous velocity of the object is the rate of change of displacement as a function of time. The velocity is distinct from the instantaneous speed which is the time rate of change of the distance traveled along a specific path.

The velocity may be equivalently defined as the time rate of change of the position vector. If one considers a moving initial position or equivalently a moving origin, then it may be referred to as a relative velocity as opposed to an absolute velocity, which is computed with respect to a point which is considered to be 'fixed in space'.

Rigid Body

In dealing with the motion of a rigid body, the term displacement can also include the rotations of the body. In this case, the displacement of a particle of the body is called linear displacement, while the rotation of the body is called angular displacement

$$\text{Average velocity} = \frac{\text{Total distance travelled in a particuller direction}}{\text{Total time taken}}$$

5.1.1 Average Velocity

Velocity is defined as the rate of change of position of an object. It is a vector quantity because it requires both magnitude and direction to define.

Average velocity is defined as the ratio of total displacement to total time.

If 's' is total the displacement of the body and 't' is the total time taken by the body to complete the displacement, we have $V_{ave} = s/t$.

5.1.2 Instantaneous Velocity

The instantaneous velocity of an object is the velocity at a certain instant of time. Velocity is the change in position divided by the change in time and the instantaneous velocity is the limit of velocity as the change in time approaches zero.

This is equivalent to the derivative of position with respect to time. Instantaneous velocity is a vector and so it has a magnitude (a value) and a direction. The unit for instantaneous velocity is meters per second (m/s).

$$\text{Instantaneous velocity} = \text{limit as change in time goes to zero}\left(\frac{\text{change in position}}{\text{change in time}}\right)$$

$$\vec{o} = \lim_{\Delta t \to o}\left(\frac{\Delta \vec{r}}{\Delta t}\right) = \frac{d\vec{r}}{dt} = \text{derivative of position with respect to time}$$

$\vec{\upsilon}$ = Instantaneous velocity (m/s)

$\Delta \vec{r}$ = Vector change in position (m)

Δt = Change in time (s)

$\frac{d\vec{r}}{dt}$ = Derivative of vector position with respect to time (m/s)

5.1.3 Speed and Acceleration

Speed is a scalar quantity that defines, as how fast an object is moving. Speed can be thought of, as the rate at which an object covers the distance. A fast-moving object has a higher speed and covers a relatively large distance in a short amount of time.

Contrast to this a slow-moving object that has a low speed; it covers a relatively small amount of distance in the same amount of time. An object without movement has a zero speed.

Acceleration

Acceleration is a vector quantity which shows the direction and magnitude of changes in velocity. Its standard units are meters per second squared or meters per second.

Instantaneous acceleration is the rate and direction at which the velocity of an object is changing at one particular moment.

5.1.4 Average Acceleration

Average acceleration is the total change in velocity over some extended period of time divided by the duration of that period.

Changing velocity means an acceleration is present.

Average Acceleration, $\bar{a}$

$$\bar{a} = \frac{\text{change in velocity}}{\text{Time Intervel}}$$

$$\bar{a} = \frac{\Delta v}{\Delta t} = \frac{V_f - v_o}{t_f - t_o}$$

Units are m/s^2 (SI), cm/s^2 (CGS) and ft/s^2 (US)

5.1.5 Variable Acceleration

A body is said to be moving with variable acceleration if its average acceleration is different between different points along its path, either in direction or magnitude or both in direction as well as magnitude.

5.1.6 Acceleration Due to Gravity

If a body is dropped from a certain height, it will accelerate because of gravity. The acceleration caused by gravity is written as "g" and is generally taken to be 9.8 ms^{-2}.

Problem

1. A ball is dropped from the Leaning Tower of Pisa, at a height of 50 m from the ground. The ball is dropped from rest and falls freely under gravity. How long will it take for the ball to hit the ground?

Solution:

Given:

$s = 50, a = 9.8, u = 0$

To find:

Time (t)

The equation connecting these four is $s = ut + \frac{1}{2} at^2$

So,

$50 = 0 + \frac{1}{2} \times 9.8 \times t^2$ Rearranging,

$t = 10.20408t = 3.19$...The time taken is 3.19s.

5.1.7 Newton's Laws of Motion

Every object in a state of uniform motion tends to remain in that state of motion unless an external force is applied to it.

Newton's Second Law

The relationship between an object's mass m, its acceleration a and the applied force F is F = ma. Acceleration and force are vectors. According to this law the direction of the force vector is the same as the direction of the acceleration vector.

According to Newton, a force causes only a change in velocity (an acceleration) it does not maintain the velocity.

Newton's Third Law

For every action, there is an equal and opposite reaction.

5.2 Rectilinear Motion

Rectilinear motion is also called as straight-line motion. This type of motion describes the movement of a body or particle.

A body is said to experience rectilinear motion if any two particles of the body travel the same distance along two parallel straight lines. The figures below illustrates rectilinear motion for a body or particle.

Rectilinear Motion for a Particle

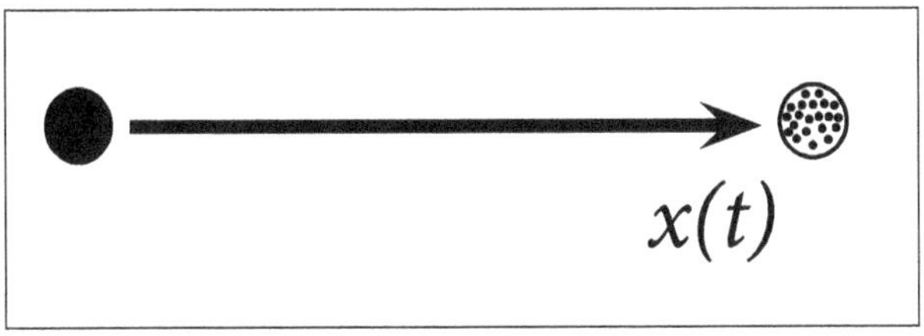

Rectilinear Motion for a Body

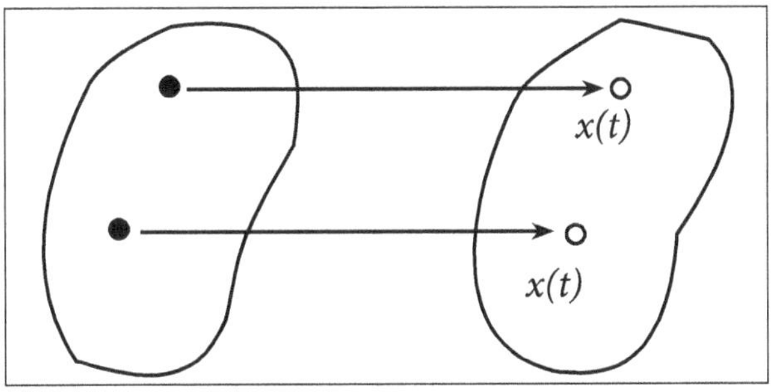

In the above figures, x(t) represents the position of the particles along the direction of motion, as a function of time t.

Given the position of the particles, x(t), we can calculate the acceleration, displacement and velocity. These are important quantities to consider when evaluating the problems in kinematics. A general assumption, which applies to numerous problems involving rectilinear motion, is that acceleration is constant. With acceleration as a constant we can derive equations for the position, displacement and velocity of a body or particle, experiencing rectilinear motion.

The easiest way to derive these equations is by using Calculus.

The acceleration is given by, $\frac{d^2x}{dt^2} = a$

Where a is the acceleration, which is a constant.

Integrate the above equation with respect to time, to obtain velocity. This gives us, $v(t) = \int a\,dt = C_1 + at$ Where,

C_1 is a constant

v(t) is the velocity

Integrate the above equation with respect to time, to obtain position. This gives us,

$$x(t)=\int v(t)dt=C_2+C_1t+\frac{1}{2}at^2$$

Where C_2 is a constant and x(t) is the position.

The constants C_1 and C_2 are found out by using the initial conditions at time t = 0.

The initial conditions are,

At time t = 0 the position is x_1.

At time t = 0 the velocity is v_1.

Substituting these two initial conditions into the above two equations we get,

$$\upsilon(0)=\upsilon_1=C_1$$
$$x(0)=x_1=C_2$$

Therefore $C_1 = v_1$ and $C_2 = x_1$.

$$x(t)=x_1+v_1t+\frac{1}{2}at^2$$

$\upsilon(t)=\upsilon_1+$ at For convenience, set $x(t)=x_2$ and $v(t)=v_2$.

As a result, $x_2=x_1+v_1t+\frac{1}{2}at^2 \rightarrow$ Position equation ...(1)

$\upsilon_2=\upsilon_1+$ at $\rightarrow$ Velocity equation ...(2)

Displacement is defined as $\Delta d=x_2-x_1$.

Therefore, equation (1) becomes, $\Delta d=v_1t+\frac{1}{2}at^2 \rightarrow$ Displacement equation...(3)If we wish to find an equation that doesn't involve time t we can combine equations (2) and (3) to eliminate time as a variable. This gives us,

$\upsilon_2^2=\upsilon_1^2+2a(\Delta d)\rightarrow 2^{nd}$ Velocity equation...(4) Equations (1), (2), (3) and (4) describes the motion of bodies or particles experiencing rectilinear motion, where acceleration a is constant.

For the cases where acceleration is not a constant, new expressions have to be derived for the position, displacement and velocity of a particle. If the acceleration is known as a function of time, we can use Calculus to find the displacement, velocity and position in the same manner as before.

Alternatively, if we are given the position x(t) as a function of time, we determine the velocity by differentiating x(t) once and we determine the acceleration by differentiating x(t) twice.

For instance, let's say the position x(t) of a particle is given by,

$(t)=\cos(2t)+4t^3$ Thus, the velocity v(t) is given by, $v(t)=\frac{dx}{dt}=-2\sin(2t)+12t^2$ The acceleration a(t) is given by, $a(t)=\frac{d^2x}{dt^2}=-4\cos(2t)+24t$

5.2.1 Numerical Problems

1. A car is driven along a straight track with position given by s(t) = 150t – 300 ft (t in seconds). (a) Let us determine v(t) and a(t).

Solution:

Given:

$$s(t) = 150t - 300 \text{ ft}$$

To find:

v(t) and a(t)

So,

$$v(t)=s'(t)=150 \text{ ft/s and } a(t)=v'(t)=0 \text{ ft/s}^2$$

Use calculus to find the displacement and total distance travelled over the interval [1, 4].

The displacement on [1, 4] is simply the definite integral of velocity on [1, 4].

$$\text{Displacement} = \int_1^4 150\, dt = (150t)\Big|_1^4 = 150.4-150.1=450\,\text{ft}$$

Since the velocity of the car is constant, the car is always moving in the same direction. Therefore, the total distance traveled is the same as the displacement in this case.

2. A projectile is fired upward from a 15.3 m cliff at a speed of 19.6 m/s and allowed to fall into a valley below. The acceleration g due to Earth's gravity is about 9.8 m/s² or about 32 ft/s², downward. Let us determine the velocity, maximum height of projectile, displacement and total distance traveled.

Solution:

Given:

$$a(t) = -9.8 \text{ m/s}^2$$

To find:

Velocity, maximum height of projectile, displacement and total distance traveled.

v(t) is used to find the time at which the projectile reaches its maximum height. The maximum

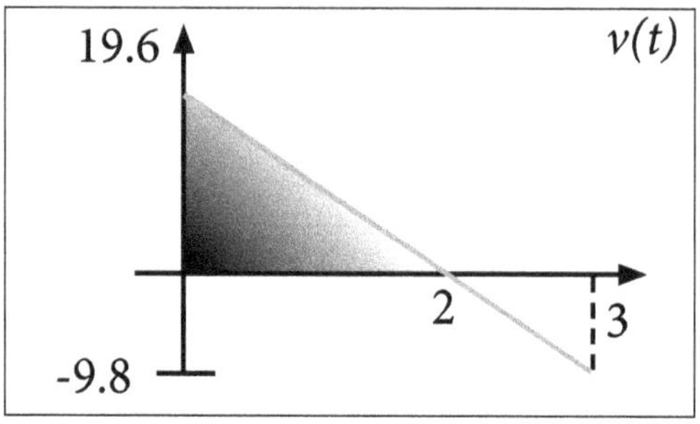

If a(t) = – 9.8, then v(t) = – 9.8t + C for some constant C.

Note that v(0) = C in this problem, so that C is the initial velocity.

Therefore, v(t) = −9.8t + 19.6 m/s.

The maximum height occurs when the velocity is zero, so -9.8t + 19.6 = 0 implies that the maximum height occurs at t = 2 seconds.

Although we do not have a position function, we can find the maximum height using geometry.

Since the maximum height in this problem is simply the displacement over the first 2 seconds and the displacement is the net area bounded by the velocity curve, we see that the maximum height is the area of the shaded triangle, which is (1/2)(2)(19.6) = 19.6 meters.

The displacement and total distance traveled over the interval [0, 3] can be found out by using geometry.

The displacement is the net area bounded by v(t) and the total distance traveled is the total area. In this case, we can use the two triangles in the figure to compute displacement on [0,3] as (1/2)(2)(19.6) - (1/2)(1)(9.8) = 14.7 meters and the total distance traveled on [0,3] as (1/2)(2)(19.6) + (1/2)(1)(9.8) = 24.5 meters.

5.3 Curvilinear Motion

In a plane, curvilinear motion is motion along a plane curve. The velocity and acceleration of a point on such a curve can be expressed either as,

- Rectangular components.

- Tangential and normal components.
- Radial and transverse components.

Rectangular Components

The position vector r(t), the velocity vector v(t) and the acceleration vector a(t) are given by,

$$r(t)=x(t)i+y(t)j+z(t)k$$

$$\upsilon(t)=\upsilon_x(t)\imath+\upsilon_y(t)j+\upsilon_z(t)k=\dot{x}(t)i+\dot{y}(t)j+\dot{z}(t)k$$

$$a(t)=a_x(t)i+a_y(t)j+a_z(t)k=\ddot{x}(t)i+\ddot{y}(t)j+\ddot{z}(t)k$$

Where the over-dot represents time differentiation.

The rectangular components of velocity and acceleration are given by,

$$\upsilon_x(t)=\dot{x}(t),\upsilon_y(t)=\dot{y}(t),\upsilon_z(t)=(t)\dot{z}(t)$$

$$a_x(t)=\ddot{x}(t),a_y(t)=\ddot{y}(t),a_z(t)=(t)\ddot{z}(t)$$

The components of position are given by,

$$x(t)=x(t_1)+\int_{t_1}^{t}v_x(s)ds$$

$$y(t)=y(t_1)+\int_{t_1}^{t}v_y(s)ds$$

$$z(t)=z(t_1)+\int_{t_1}^{t}v_z(s)ds$$

The components of velocity are given by,

$$v_x(t)=v_x(t_1)+\int_{t_1}^{t}a_x(s)ds$$

$$v_y(t)=v_y(t_1)+\int_{t_1}^{t}a_y(s)ds$$

$$v_z(t)=v_z(t_1)+\int_{t_1}^{t}a_z(s)ds$$

Also,

$$v_x^2(x) = v_x^2(x_1) + 2\int_{x_1}^{x} a_x(x)dx$$

Tangential and Normal Components

Referring to below figure, the velocity vector v can be written as,

Where,

v = Magnitude of the velocity vector

n_t = Tangential unit vector directed along the velocity vector

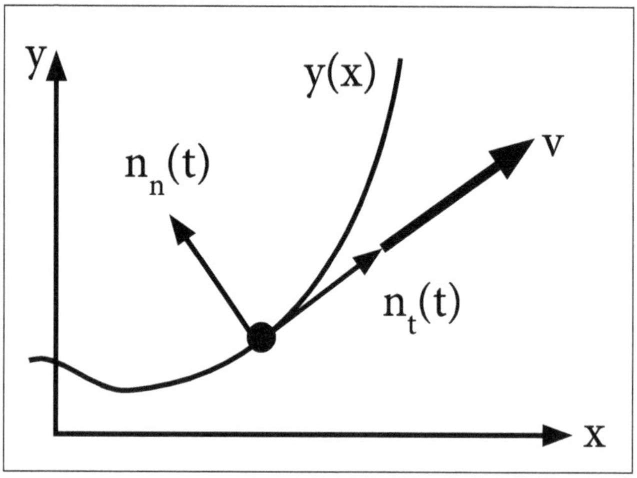

Also,

$$\dot{n}_t = \dot{\theta} n_n$$

And,

$$\dot{n}_n = -\dot{\theta} n_f$$

Where n_t = The normal unit vector defined to be perpendicular to the tangential unit vector.

In figure below, the radius of curvature p of the path at time t is shown which is obtained by the intersection of the lines extending from $n_t(t)$ and $n_t(t + \Delta t)$ where Δt is time increment. The angle θ changes an incremental amount $\Delta\theta$ and the point moves an incremental amount Δs.

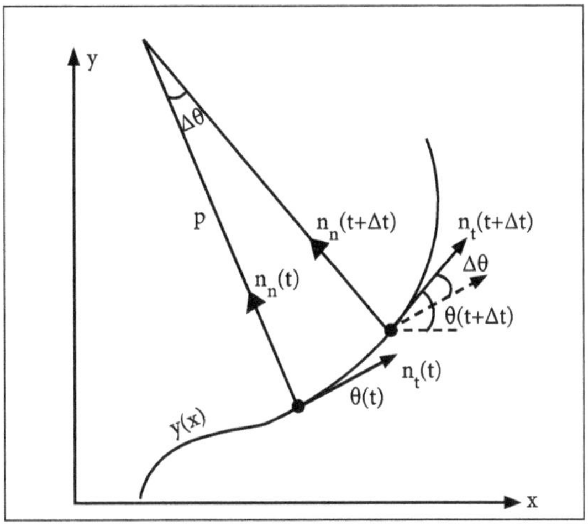

Now,

$$\rho\Delta\theta = \Delta s$$

or,

$$\rho\dot{\theta} = v$$

Where,

$$\dot{\theta} = \frac{d\theta}{dt}$$

$$v = \frac{ds}{dt}$$

Equation (1) with respect to time gives,

$$a = a_t n_t + a n_n$$

Where,

$$a_t = \dot{v}$$

$$a_n = \frac{v^2}{\rho}$$

Radial and Transverse Components

The polar form of a position vector is,

$$r = r n_r \qquad ...(2)$$

Where,

$r = |r|$ is the magnitude of r

n_r = Unit vector in the direction of r

n_t = The radial unit vector

n_θ = The circumferential unit vector

The Circumferential Unit Vector is perpendicular to n_r. Hence nt and nθ are related to i and j in figure.

$$n_r t = \cos\theta(t)i + \sin\theta(t)j$$

$$n_\theta t = -\sin\theta(t)i + \cos\theta(t)j \quad ...(3)$$

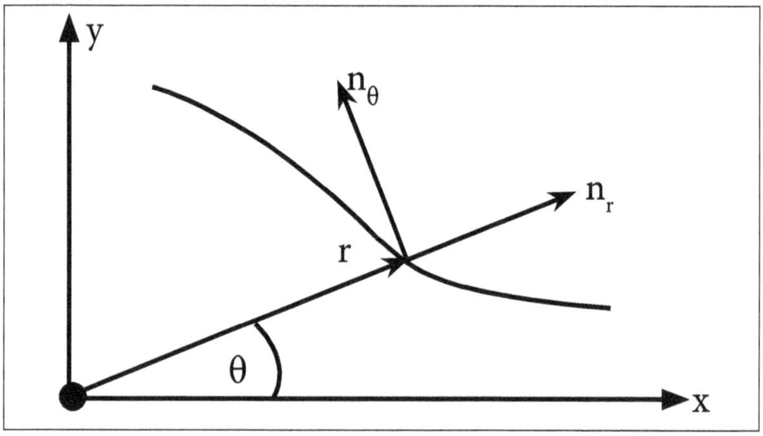

Differentiating equation (2) with respect to time gives,

$$\dot{n}_r = -\dot{\theta}\sin\theta i + \dot{\theta}\cos\theta j = \dot{\theta}(-\sin\theta i + \cos\theta j) = \dot{\theta} n_\theta$$

And,

$$\dot{n}_\theta(t) = -\dot{\theta} n_r$$

Hence the derivatives of the unit vectors are given by,

$$\dot{n}_r(t) = -\dot{\theta} n_\theta$$

$$\dot{n}_\theta(t) = -\dot{\theta} n_r \quad ...(4)$$

Similarly from equations 2 and 4, we get

$$v = v_r n_r + v_\theta n_\theta \quad \ldots(5)$$

Where,

$$\upsilon_r = \dot{r}$$

$$\upsilon_\theta = r\dot{\theta}$$

Differentiating equation (5) with respect to time gives,

$$a = a_r n_r + a_\theta n_\theta$$

Where,

$$a_r = \ddot{r} - r\dot{\theta}^2$$

$$a_\theta = r\ddot{\theta} + 2\dot{r}\dot{\theta}$$

5.3.1 Super Elevation

Super elevation may be defined as the raising of the outer edge of the road along a curve in order to reduce or neutralize the effects of centrifugal force, that acts on the vehicle and in turn the stability of the vehicle is disturbed.

Whenever the vehicle is moving in a curved path, it is acted upon by the force outward the curve known as centrifugal force. This force is always acting in the horizontal direction through the center of gravity of the vehicle and always perpendicular to the weight of the vehicle.

If centrifugal force is greater than the force between the road and the tyres, then the vehicle will skid outward. If centrifugal force is less than the force between the road and the tyres, then the vehicle will overturn to the inside of the curve.

Super elevation is to be provided in case of curves only. If the road surface is plain or flat i.e. there is no curve, then the tendency of the centrifugal force is through the vehicle from the center line and the only force which will prevent the vehicle is frictional force.

This force depends upon centripetal force and the weight of the vehicle. The procedure in which outer edge is raised with respect to the inner edge is known as super elevation. Its units are meter/meter or feet/feet.

Provision of Super Elevation in the Field

- For the provision of super elevation in the field, we may provide rotation about the inner edge, when the cutting of the road is required.

- Similarly, rotation about outer edge is preferred, when the filling of the road is required.
- For the good appearance, rotation about outer edge is preferred.
- Rotation about center is preferred when inner and outer edges are at same level.

5.3.2 Projectile Motion

We often experience many kinds of motions in our daily life. Projectile motion is one among them. A projectile is some object thrown in air or space. Curved path along which the projectile travels is termed as trajectory.

Projectile Motion is the free fall motion of any body in a horizontal path with constant velocity.

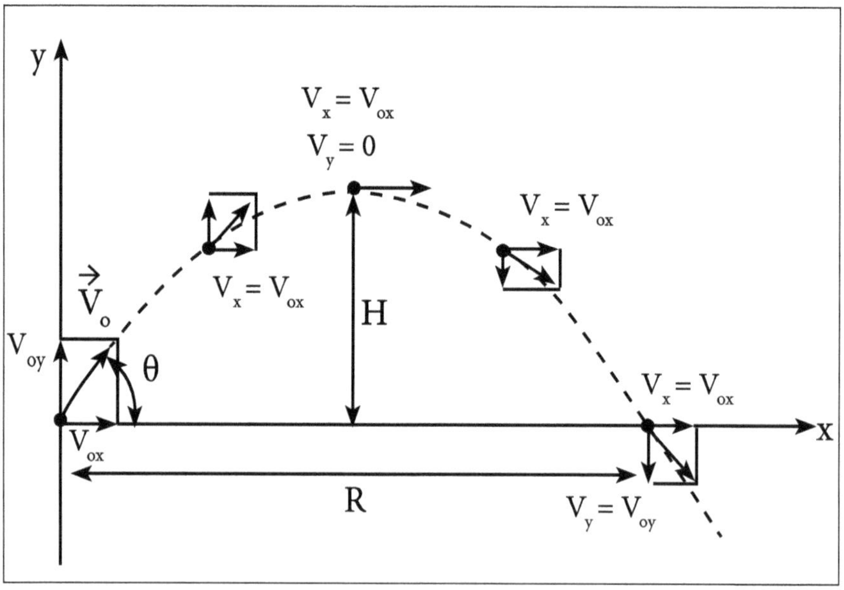

Projectile Motion Formula (trajectory formula) is given by,

Horizontal distance, $x = V_{xt}$

Horizontal velocity, $V_x = V_{xo}$

Vertical distance, $y = v_{yo}t - \frac{1}{2}gt^2$

Vertical velocity, $v_y = v_{yo} - gt$ Where, V_x = The velocity along $x-$axis, V_{xo} = The initial velocity along $x-$axis, V_y = The velocity along $y-$axis, V_{yo} = The initial velocity along y-axis. g = The acceleration due to gravity and t = Time taken. Equations related to trajectory motion are given by,

$$\text{Time of flight, } t = \frac{2v_0 \sin\theta}{g}$$

$$\text{Maximum height reached, } H = \frac{v_0^2 \sin^2\theta}{2g}$$

$$\text{Horizontal range, } R = \frac{v_0^2 \sin 2\theta}{g}$$

Where, V_0 – The initial Velocity, sin θ - The component along y-axis, cos θ - The component along x-axis. Projectile Motion formula is used to find the distance, velocity and time taken in the projectile motion.

5.3.3 Relative Motion

Generally, a moving body is observed by a person who is at rest. Considering the observers position at rest we are developing fixed axis reference. Such a set of fixed axes is defined as absolute or Newtonian or inertial frame of reference. For most of the moving bodies, the Earth is regarded as fixed although Earth itself is moving in space. Motion referred with such fixed axis is called an absolute motion.

However, if the axes reference is attached to a moving object then such motion is termed as relative motion. It means person in moving object is observing another object which is also in motion.

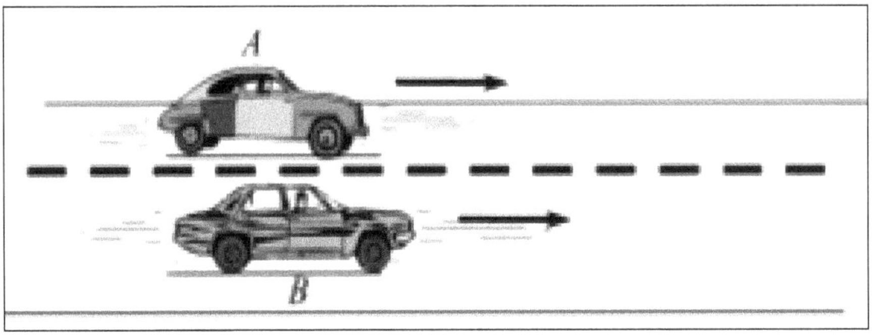

Cars A and B are moving in the same direction on road parallel to each other. Car A is moving with a speed of 60 km/hr and car B is moving with 80 km/hr. Car A in relation to car B is moving backward with speed 20 km/hr whereas car B in relation to car A is moving forward with speed of 20 km/hr.

Observation of drivers of car *A* and car *B* with respect to each is developing relative motion between them.

Relative Motion between Two Particles

Consider two particles A and B moving on different paths as shown in figure. Here xOy is the fixed frame of reference. Therefore, absolute position of A is given by $r_A = OA$ and of B is $r_B = OB$. Therefore, relative position of B with respect to A is written as $r_{B/A}$

By triangle law of vector addition, we have,

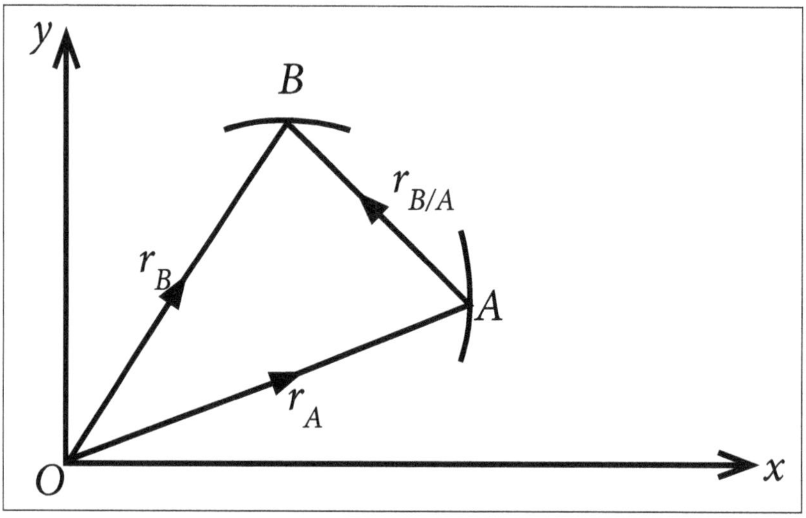

$$r_A + r_{B/A} = r_B$$

$\therefore$ Relative position of B with respect to A, $r_{B/A} = r_B - r_A$ ...(1)

Differentiating equation (1) with respect to t, we have

Relative velocity of B with respect to A $v_{B/A} = v_B - v_A$ ...(2)

Further Differentiating equation (2) with respect to t, we have

Relative direction of B with respect to A $a_{B/A} = a_B - a_A$ ...(3)

5.3.4 Numerical Problems

1. A train covers 60 miles between 2 p.m. and 4 p.m. Let us determine the speed at 3 p.m.

Solution:

Given:

Distance = 60 miles

To find:

Speed:

The speed is calculated by dividing travelled distance (60 miles) and travelled time (4pm – 2pm = $\frac{60 \text{ miles}}{2 \text{ hours}} = 30$ mph 2hours).

2. Is it possible that the car could have accelerated to 55mph within 268 meters. Let us determine the acceleration of the car from 0 to 60 mph in 15 seconds.

Solution:

Given:

Speed = 55 mph

Distance = 268 meters

To find:

Let us find the maximum acceleration of the car,

The car can accelerate from 0 to 60 mph = 60*0.447 m/s = 26.8 m/s in 15 seconds. Then maximum acceleration is,

$$\frac{26.8}{15} = 1.8 \text{m/s}^2$$

If the car needs to accelerate to 55 mph = 55*0.447 m/s = 24.6 m/s within 268 meters then its acceleration should be,

$$a = \frac{24.6^2}{2 * 268} = 1.13 \text{m/s}^2$$

This acceleration is less than the maximum possible acceleration, so the car can reach the speed of 55 mph within 268 meters.

5.4 Motion Under Gravity

When we throw an object upwards, it will eventually fall back to the ground under the earth's gravity. In fact all the objects near the earth's surface fall with a constant acceleration of about 9.8 ms^{-2}. This is called the acceleration due to gravity and is usually represented by the symbol g.

An object that is thrown vertically upwards decelerates under the earth's gravity. Its speed decreases till it attains a maximum height, where the velocity is zero. Then it is accelerated uniformly downwards under gravity. When it returns to the point of projection, it has the same speed as that, at the instant of projection. In addition, the duration of the upward motion is exactly equal to that of the downward motion.

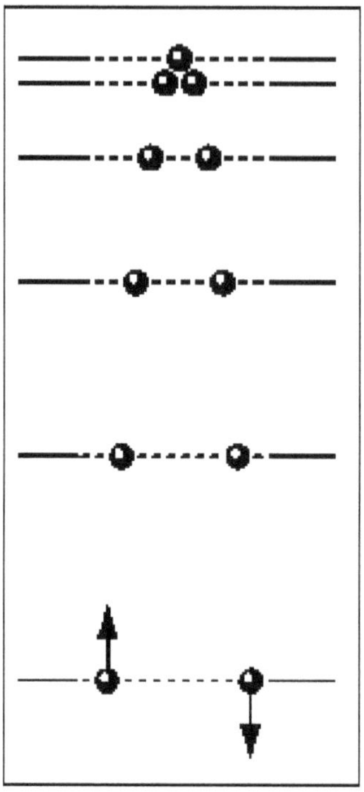

5.4.1 Numerical Problems

1. A tennis ball is thrown vertically upward at an initial velocity of 3 ms^{-1}. Let us determine the maximum height that it can reach. How long would it take to reach this height?

Solution:

Given:

Initial velocity = 3 ms^{-1}

To find:

Maximum height

Time to reach the maximum height

Firstly, we have to define the sign convention. Usually we take the upward direction as positive. Hence the initial velocity is $u = +\,3ms^{-1}$ and the acceleration under gravity is $a = g = -\,10\ ms^{-2}$.

At the highest point, the velocity $v = 0$.

Apply $v^2 - u^2 = 2as$, we have,

$$0 - 3^2 = 2 \times (-10) \times s$$

$$S = 0.45 \text{ m}$$

To find the time t, we apply,

$$v - u = at$$

$$0 - 3 = -10t$$

$$t = 0.3 \text{ s}$$

2. During an explosion, a piece of the bomb is projected vertically upwards at a velocity of 25 ms^{-1}. Let us determine how long would it take to fall back to the ground?

Solution:

Given:

$$U = 25 \text{ ms}^{-1},\ a = g = -10 \text{ ms}^{-2},\ s = 0 \text{ (Since it returns to the ground)}$$

To find:

Time (t)

Applying $s = ut + \frac{1}{2}at^2$, we have,

$$0 = 25t + \frac{1}{2}(-10)t^2$$

$$10t^2 - 50t = 0$$

This is a quadratic equation in t with two solutions.

$$t(10t - 50) = 0$$

$$t = 0 \text{ or } t = 50/10 = 5.0 \text{ s}$$

We may wonder why there are two solutions of t. In fact the first solution $t = 0$ corresponds to the instant of projection. The solution corresponding to the duration of flight should be $t = 5s$.

Permissions

All chapters in this book are published with permission under the Creative Commons Attribution Share Alike License or equivalent. Every chapter published in this book has been scrutinized by our experts. Their significance has been extensively debated. The topics covered herein carry significant information for a comprehensive understanding. They may even be implemented as practical applications or may be referred to as a beginning point for further studies.

We would like to thank the editorial team for lending their expertise to make the book truly unique. They have played a crucial role in the development of this book. Without their invaluable contributions this book wouldn't have been possible. They have made vital efforts to compile up to date information on the varied aspects of this subject to make this book a valuable addition to the collection of many professionals and students.

This book was conceptualized with the vision of imparting up-to-date and integrated information in this field. To ensure the same, a matchless editorial board was set up. Every individual on the board went through rigorous rounds of assessment to prove their worth. After which they invested a large part of their time researching and compiling the most relevant data for our readers.

The editorial board has been involved in producing this book since its inception. They have spent rigorous hours researching and exploring the diverse topics which have resulted in the successful publishing of this book. They have passed on their knowledge of decades through this book. To expedite this challenging task, the publisher supported the team at every step. A small team of assistant editors was also appointed to further simplify the editing procedure and attain best results for the readers.

Apart from the editorial board, the designing team has also invested a significant amount of their time in understanding the subject and creating the most relevant covers. They scrutinized every image to scout for the most suitable representation of the subject and create an appropriate cover for the book.

The publishing team has been an ardent support to the editorial, designing and production team. Their endless efforts to recruit the best for this project, has resulted in the accomplishment of this book. They are a veteran in the field of academics and their pool of knowledge is as vast as their experience in printing. Their expertise and guidance has proved useful at every step. Their uncompromising quality standards have made this book an exceptional effort. Their encouragement from time to time has been an inspiration for everyone.

The publisher and the editorial board hope that this book will prove to be a valuable piece of knowledge for students, practitioners and scholars across the globe.

Index

Printed in the USA
CPSIA information can be obtained
at www.ICGtesting.com
LVHW080735200324
774517LV00074B/717

9 781639 897131